METHODS IN ECOLOGY

Ecological Data: Design, Management and Processing

EDITED BY

WILLIAM K. MICHENER
J.W. Jones Ecological Research Center
Route 2, Box 2324
Newton, GA 31770

AND

JAMES W. BRUNT
LTER Network Office
Department of Biology
University of New Mexico
Albuquerque, NM 87131-1091

b
Blackwell
Science

© 2000 by
Blackwell Science Ltd
Editorial Offices:
Osney Mead, Oxford OX2 0EL
25 John Street, London WC1N 2BL
23 Ainslie Place, Edinburgh EH3 6AJ
350 Main Street, Malden
 MA 02148-5018, USA
54 University Street, Carlton
 Victoria 3053, Australia
10, rue Casimir Delavigne
 75006 Paris, France

Other Editorial Offices:
Blackwell Wissenschafts-Verlag GmbH
Kurfürstendamm 57
10707 Berlin, Germany

Blackwell Science KK
MG Kodenmacho Building
7–10 Kodenmacho Nihombashi
Chuo-ku, Tokyo 104, Japan

The right of the Author to be
identified as the Author of this Work
has been asserted in accordance
with the Copyright, Designs and
Patents Act 1988.

All rights reserved. No part of
this publication may be reproduced,
stored in a retrieval system, or
transmitted, in any form or by any
means, electronic, mechanical,
photocopying, recording or otherwise,
except as permitted by the UK
Copyright, Designs and Patents Act
1988, without the prior permission
of the copyright owner.

First published 2000

Set by Graphicraft Limited, Hong Kong
Printed and bound in Great Britain
at the Alden Press, Oxford and Northampton.

The Blackwell Science logo is a
trade mark of Blackwell Science Ltd,
registered at the United Kingdom
Trade Marks Registry

DISTRIBUTORS

Marston Book Services Ltd
PO Box 269
Abingdon, Oxon OX14 4YN
(*Orders*: Tel: 01235 465500
 Fax: 01235 465555)

USA
Blackwell Science, Inc.
Commerce Place
350 Main Street
Malden, MA 02148-5018
(*Orders*: Tel: 800 759 6102
 781 388 8250
 Fax: 781 388 8255)

Canada
Login Brothers Book Company
324 Saulteaux Crescent
Winnipeg, Manitoba R3J 3T2
(*Orders*: Tel: 204 837 2987)

Australia
Blackwell Science Pty Ltd
54 University Street
Carlton, Victoria 3053
(*Orders*: Tel: 3 9347 0300
 Fax: 3 9347 5001)

A catalogue record for this title
is available from the British Library

ISBN 0-682-05231-7

Library of Congress
Cataloging-in-publication Data

Ecological data : design, management, and
processing / edited by William K. Michener
and James W. Brunt.
 p. cm.
 This work resulted from two workshops and a
 working group.
 Includes bibliographical references.
 ISBN 0-632-05231-7
 1. Ecology–Data processing.
 I. Michener, William K.
 II. Brunt, James W.
QH541.15.E45 E24 2000
577'0285–dc21   99-042236

For further information on
Blackwell Science, visit our website:
www.blackwell-science.com

ECOLOGICAL DATA:
DESIGN, MANAGEMENT
AND PROCESSING

0195785

1 Week Loan

This book is due for return on or before the last date shown below

University of Cumbria
24/7 renewals Tel:0845 602 6124

METHODS IN ECOLOGY

Series Editors
J.H. LAWTON FRS
Imperial College at Silwood Park
Ascot, UK

G.E. LIKENS
Institute of Ecosystem Studies
Millbrook, USA

Contents

Contributors, vii

The Methods in Ecology Series, viii

Preface, ix

Acknowledgements, xi

1 Research Design: Translating Ideas to Data, 1
   William K. Michener

2 Data Management Principles, Implementation and Administration, 25
   James W. Brunt

3 Scientific Databases, 48
   John H. Porter

4 Data Quality Assurance, 70
   Don Edwards

5 Metadata, 92
   William K. Michener

6 Archiving Ecological Data and Information, 117
   Richard J. Olson & Raymond A. McCord

7 Transforming Data into Information and Knowledge, 142
   William K. Michener

8 Ecological Knowledge and Future Data Challenges, 162
   William K. Michener

Index, 175

Dedicated to:

James T. (Tom) Callahan
*friend, ecologist, and visionary who has long
championed data management, ecosystem science,
long-term ecological research, and biological
field stations and marine laboratories.*

Contributors

James W. Brunt
LTER Network Office
Department of Biology
University of New Mexico
Albuquerque, NM 87131-1091

Don Edwards
Department of Statistics
University of South Carolina
Columbia, SC 29208

Raymond A. McCord
Oak Ridge National Laboratory
PO Box 2008, MS-6407, Bldg. 1507
Oak Ridge, TN 37831-6407

William K. Michener
J.W. Jones Ecological Research Center
Route 2, Box 2324
Newton, GA 31770

Richard J. Olson
Oak Ridge National Laboratory
PO Box 2008, MS-6407, Bldg. 1507
Oak Ridge, TN 37831-6407

John H. Porter
Department of Environmental Sciences
Clark Hall
University of Virginia
Charlottesville, VA 22903

The Methods in Ecology Series

The explosion of new technologies has created the need for a set of concise and authoritative books to guide researchers through the wide range of methods and approaches that are available to ecologists. The aim of this series is to help graduate students and established scientists choose and employ a methodology suited to a particular problem. Each volume is not simply a recipe book, but takes a critical look at different approaches to the solution of a problem, whether in the laboratory or in the field, and whether involving the collection or the analysis of data.

Rather than reiterate established methods, authors have been encouraged to feature new technologies, often borrowed from other disciplines, that ecologists can apply to their work. Innovative techniques, properly used, can offer particularly exciting opportunities for the advancement of ecology.

Each book guides the reader through the range of methods available, letting ecologists know what they could, and could not, hope to learn by using particular methods or approaches. The underlying principles are discussed, as well as the assumptions made in using the methodology, and the potential pitfalls that could occur—the type of information usually passed on by word of mouth or learned by experience. The books also provide a source of reference to further detailed information in the literature. There can be no substitute for working in the laboratory of a real expert on a subject, but we envisage the Methods in Ecology Series as being the 'next best thing'. We hope that, by consulting these books, ecologists will learn what technologies and techniques are available, what their main advantages and disadvantages are, when and where not to use a particular method, and how to interpret the results.

Much is now expected of the science of ecology, as humankind struggles with a growing environmental crisis. Good methodology alone never solved any problem, but bad or inappropriate methodology can only make matters worse. Ecologists now have a powerful and rapidly growing set of methods and tools with which to confront fundamental problems of a theoretical and applied nature. We see the Methods in Ecology Series as a major contribution towards making these techniques known to a much wider audience.

John H. Lawton
Gene E. Likens

Preface

Ecology has experienced a computational revolution during the past quarter-century. During this period, we have seen computers shrink from room-size mainframes that required keypunched computer cards for data input to the powerful, small-footprint PCs that occupy today's desktops. Early data management was relatively straightforward, consisting of colour-coded card decks and the occasional rubber band for 'large' data sets. The advent of the calculator enabled ecologists to hand-compute solutions to regression equations; quality assurance meant punching everything twice or until identical answers were obtained.

Everything technological has changed. Sensors automatically collect massive quantities of data. Computers are faster and smaller, store tremendous volumes of data, and support analyses, graphics, and visualizations that were unthinkable even a few years ago. Technological changes have, in turn, expanded the ways we think about and do science. Who, for example, could envision landscape ecology without digital imagery acquired by satellite? All branches of ecology have taken advantage of new technologies to expand their scope of inquiry.

'Is all roses?' we might wonder. In some respects, the answer is 'yes'. For example, comprehensive scientific databases are being developed to address formerly intractable problems. Computers, databases, software, and statistical analyses are becoming increasingly powerful and sophisticated. In other respects, the answer is a resounding 'no'. Projects are completed and one or more publications are produced; yet the data are incomplete, not quality-assured, undocumented, unavailable, or non-existent. In such cases, data are treated as consumables—a means to an end—and are poorly managed during their short 'life-span'.

Despite the ecological and technological revolutions, 'improvements' have not always followed convergent pathways. Technologists (e.g. computer scientists), data analysts (e.g. statisticians), and ecologists are more specialized and it is rare that one individual is well versed in all areas. It has been suggested that ecological research groups twenty years hence will employ equal numbers of technologists and ecologists. Regardless, there is a clear trend toward increasing technological sophistication at ecological research facilities and environmental resource management agencies throughout the world.

Increased technological sophistication has had its downside. In the last decade, we have seen computers (i.e. mainframes, minicomputers, and many PCs), database management systems, operating systems, and peripherals come and go. With the high turnover in technology companies, one is frequently justified in checking the stock (financial) pages before purchasing a particular product. In addition, we have seen the implementation of high-end, expensive technology and databases at various organizations that quickly became cyber-albatrosses. At least some of the 'failures' can be attributed to the paucity of information on how best to meld the ecological and digital revolutions. The chapters in this book were written in an attempt to partially address this shortcoming.

Several factors, hopefully, set this volume apart from others in the discipline-specific technological literature. First, the chapter authors are all ecologists ranging from population and landscape ecologists, to computational and statistical ecologists. Second, the contributors average more than a quarter-century of experience apiece in designing, managing and processing data into information and knowledge. This experience has been gained in real-world academic and agency environments where funds are limiting, rapid results are of the essence, and time and people-power are consistently in short supply. Third, and probably most important, from these various experiences (including our share of both successes and mistakes), we have hopefully gleaned some keys to success that may preclude others from having to 'reinvent the wheel'.

This is an exciting time to be an ecologist. New sensors, data collection and storage devices, and analytical and statistical methods provide us with an enormously powerful toolkit to address extremely challenging and socially important ecological questions. In writing this book, we hope that others may learn from both our mistakes and insights, especially as related to managing data and information. This book makes no pretence of being a textbook on statistics, scientific databases, or experimental design. Instead, we attempt to provide the reader with a useful list of pointers to pertinent references in those fields. Finally, we hope that in some small way we have contributed to blurring the distinction between the ecological and digital revolutions.

WKM & JWB

Acknowledgements

We thank Anne Miller for her expert assistance in technical editing and Gene Likens and Ian Sherman for encouraging this endeavour. John Briggs, Daniel Morton, and Terry Parr reviewed the book and provided many thoughtful insights. Jeff Hollister and Sam Simkin assisted with figure preparation. Numerous ecologists and informatics specialists influenced the ideas that are presented in this book, including John Briggs, Tom Callahan, Walt Conley, Jerry Franklin, Jim Gosz, Bruce Hayden, John Helly, Tom Kirchner, John Pfaltz, Robert Robbins, Susan Stafford, and the LTER information managers. Additional input to various chapters came from Mike Conner, Steve Demaris, Rick Haeuber, Jeff Hollister, Steve Jack, Margaret Shannon, and Sam Simkin.

Portions of Chapters 3–6 were initially prepared for a 1997 workshop at the University of New Mexico ('Data and Information Management in the Ecological Sciences'; DIMES). The DIMES workshop was sponsored by the National Science Foundation (Grant #DBI-97-23407) and organized by William Michener, Jim Gosz, Arthur McKee and John Porter. In addition, this work resulted from two workshops ('Developing the Conceptual Basis for Restoration Biology'; 'Organization of Biological Field Stations Meeting') and a Working Group ('Interdisciplinary Synthesis of Recent Natural and Managed Floods') that were supported by the National Center for Ecological Analysis and Synthesis, a centre funded by NSF (Grant #DEB-94-21535), the University of California—Santa Barbara, the California Resources Agency, and the California Environmental Protection Agency. The Restoration Biology workshop was ably organized by Edith Allen, Wallace Covington, and Donald Falk; the proceedings of which were published in a special issue of *Restoration Ecology* (Volume 5, Number 4). Additional support was provided by the National Science Foundation (Grant #DEB-9411974 to the University of Virginia).

CHAPTER 1

Research Design: Translating Ideas to Data

WILLIAM K. MICHENER

1.1 Introduction

Most ecological data have been collected by individuals or small groups of investigators in small plots (1 m² or less) over relatively short time periods (Kareiva & Anderson 1988; Brown & Roughgarden 1990). In response to increased societal and scientific interest in issues such as global change, biodiversity and sustainability, ecologists are questioning how ecological patterns and processes vary in time and space and are developing an understanding of the causes and consequences of this variability (Levin 1992).

As the breadth and depth of questions have broadened, ecologists are expanding their scope of scientific inquiry (Fig. 1.1). Complex ecological phenomena were historically examined within the context of studies funded for 2–3 year periods; some of these phenomena are now being examined in long-term ecological research programmes that are designed to be continued

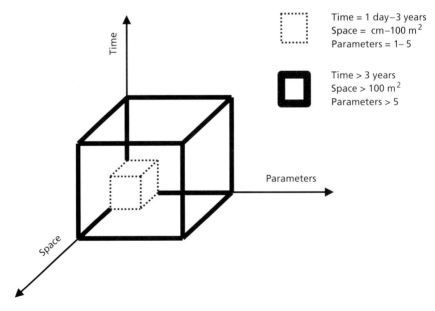

Fig. 1.1 Conceptual view of how ecologists are expanding the scope of scientific inquiry to broader spatial, temporal and thematic scales.

for decades (Likens 1989; Franklin *et al.* 1990). Studies that were once confined to small field plots at a single site or within a single ecosystem are being expanded in size and replicated across biomes. In addition, ecologists are beginning to address questions at regional, continental and global scales. The thematic focus of ecological research is also expanding. Increasingly, ecologists are studying entire communities of plants and animals (e.g. Huston 1994), as opposed to the more traditional focus on one or a few species. Furthermore, many of the questions being addressed by ecologists require data and scientific collaboration from other disciplines. Figure 1.2 illustrates how answers to many

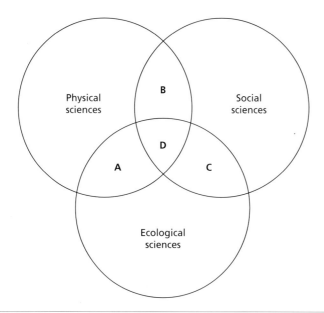

Physical Sciences—what is the interaction between flow and bank erosion, bed movement and bed load recovery?

Ecological Sciences—population recolonization rates after flooding?

Social Sciences—what current governance regimes structure management of water flow regimes and floodplains?

A Ecological-Physical Sciences—effects of flow magnitude (highs and lows) on vegetation?

B Physical-Social Sciences—where are managed floods necessary to meet societal needs?

C Ecological-Social Sciences—what are the benefits of flooding and floodplains to society (e.g., ecosystem services)?

D Ecological-Physical-Social Sciences—how do floods and flow regimes map across various geological, political and social settings (i.e. 'riverscapes')?

Fig. 1.2 Examples of scientific questions (related to management of instream flows) that arise at the boundaries among the ecological, physical and social sciences.

of the current questions of scientific interest (e.g. management of instream flows) lie at the interfaces among the ecological, physical and social sciences. In the future, it is virtually a certainty that much socially relevant ecological research will occur at the boundaries among very diverse scientific disciplines.

The new types of questions being asked by scientists have profound implications for how ecologists view and approach their scientific endeavours. The conventional approach to an ecological research project involves translating an idea or question into one or more testable hypotheses, design of an appropriate field or laboratory experiment, data collection, analysis and interpretation, and publication of the results. The entire project is frequently conceived and carried out by an individual. One of the prime characteristics of a conventional research project is that data are typically viewed as the means to an end (i.e. publication) and the data are subsequently treated as consumables. Too frequently, data are discarded or left to gather dust in the bottom of filing cabinets after the publication has been produced.

In contrast to the previous scenario, the types of questions that are envisioned to drive ecology in the future will require access to data collected by large numbers of scientists from many different disciplines under the auspices of projects which often have very diverse objectives. Good science will still continue to be performed under the single-investigator model but many of the questions being asked will require far more data than could ever be collected, managed, analysed and interpreted by a single individual. Ecology, by necessity, will increasingly place a premium on intra- and cross-disciplinary collaboration, data sharing and accessibility of high-quality, well-maintained data.

Numerous treatments of data management issues associated with global change, biodiversity and sustainability have highlighted a need for accepted protocols to assist scientists with preserving important data sets and providing guidelines for the supporting documentation necessary to interpret the data (National Research Council 1991, 1993, 1995a,b; Gosz 1994). In addition to the scientific ('question-driven') impetus that was previously outlined, funding agency mandates (e.g. National Science Foundation 1994) and the interests of various governmental agencies (e.g. see National Research Council 1993) and scientific societies (e.g. Ecological Society of America: Colwell 1995; Gross *et al.* 1995a,b) have focused increased attention on preserving, sharing and promoting the understanding of valuable data.

Questions that will challenge science in the future will require unprecedented collaboration and rapid synthesis of massive amounts of diverse data into information and, ultimately, our knowledge base. Increased emphasis on accessibility and sharing of high-quality, well-maintained, and understandable data represents a significant change in how scientists view and treat data. Despite their importance, these issues are typically not addressed in database, ecological, and statistical textbooks. For these reasons, this book focuses on ecological data, their management, and transformation into information and knowledge from both scientific and technological (i.e. data centric) viewpoints. Both the types of, and approaches to, the design of ecological studies that give rise to data are discussed in the remainder of this chapter. However, before different research approaches are presented, it may be instructive to step back and briefly examine ecological worldviews and the emergence of the questions and hypotheses that shape research.

1.2 Formulating questions and hypotheses

1.2.1 Paradigms and conceptual models

Paradigms represent the underlying worldview of ecologists and encompass prevailing beliefs, assumptions and approaches (see Pickett *et al.* 1994 for comprehensive discussion). Paradigms exist within the various sub-disciplines of ecology (e.g. the population, community and ecosystem). Two particularly important, competing paradigms that have had significant implications for ecology are the balance of nature and flux of nature. The balance of nature is a classical ecological paradigm and has served as the foundation for much of the ecological research performed in the twentieth century. Under the balance of nature, ecological systems are viewed as closed, as self-regulating, with fixed equilibrium points; succession is fixed, disturbance is exceptional and humans are excluded from ecological systems (Pickett & Ostfield 1995). In contrast, the flux of nature is a relatively new paradigm that views ecological systems as being open, as externally limited, with stable point equilibrium as a rare state; succession is rarely deterministic, disturbance is a common event and human influences on ecological systems are pervasive (Pimm 1991; Pickett *et al.* 1992; Pickett & Ostfield 1995). Clearly, the scientific approaches employed and the underlying questions and hypotheses addressed will differ depending upon the paradigms to which one subscribes.

A scientist's view of natural and human-modified environments is based on numerous presuppositions (derived from one or more paradigms) that serve as the framework from which conceptual models of the environment arise, hypotheses are formulated and studies are designed. Conceptual models underlie all ecological research. Models may be implicit or explicit and qualitative or quantitative. In essence, models represent a simplification of reality, focusing on one or more states of an ecological system, one or more system components and various relationships and interactions. Pickett *et al.* (1994) suggest that 'in a sense, models are the explicit workings out of the notions, assumptions, or concepts mentioned earlier, or they may be derived from confirmed generalizations and laws that arise as the theory becomes better developed'.

1.2.2 Hypothesis generation

Expectations (or predictions about unknown events) can be derived from any particular conceptual model. These expectations can then be formulated as testable statements (i.e. hypotheses). Underwood (1997) depicts hypothesis formulation as an iterative process whereby an hypothesis (a prediction based on a model) is restated as its logical opposite (null hypothesis) (Fig. 1.3). An experiment is then designed and performed to test the null hypothesis, which is either rejected (lending credence to the hypothesis and model) or accepted (implying that the underlying model may be deficient). Importantly, as noted by Underwood (1997), the process is continuous; regardless of whether the null hypothesis is rejected or accepted, interpretation should lead to new hypotheses and experimental tests (see also Popper 1968; Kuhn 1970).

Bernstein and Goldfarb (1995) have proposed a useful approach for generating and evaluating ecological hypotheses based on explicitly defining the presuppositions of the observer (i.e. ecologist). They offer six characteristic, basic presuppositions:

1. effect or type of causality (e.g. correlation, feedback loops);
2. components or organisms and habitat features (e.g. habitat, population, ecosystem, landscape);
3. operations or ways the components relate (e.g. sedimentation, competition);
4. spatial scale;
5. evidence or classes of valid data (e.g. descriptive survey, field experiment) and
6. temporal scale.

Several other factors may also warrant consideration, particularly for studies that will be conducted in the field. For instance, site history and the historical range of variability in ecological structure and function for an area may affect our presuppositions about the environment, and may predispose the outcome of a research project (Pickett 1989; Morgan *et al*. 1994). For in-the-field research projects, characteristics of the broader encompassing ecosystem or landscape, as well as spatial interconnectivity and other landscape attributes may need to be considered, particularly in relation to how they affect processes like material flow (e.g. water) and animal movement across the study area (Turner & Gardner 1991; Forman 1995).

Consideration and explicit documentation of the parameters outlined above could facilitate research design, interpretation of results from a specific study, and comparison of results from multiple studies. Preliminary thought exercises involving the alteration of individual presuppositions, as suggested by Bernstein and Goldfarb (1995), could also lead to the formulation of more robust ecological hypotheses. Allen and Hoekstra (1992), for example, contend that when defining observation levels (or study components as defined in Bernstein & Goldfarb 1995), more comprehensive understanding and robust

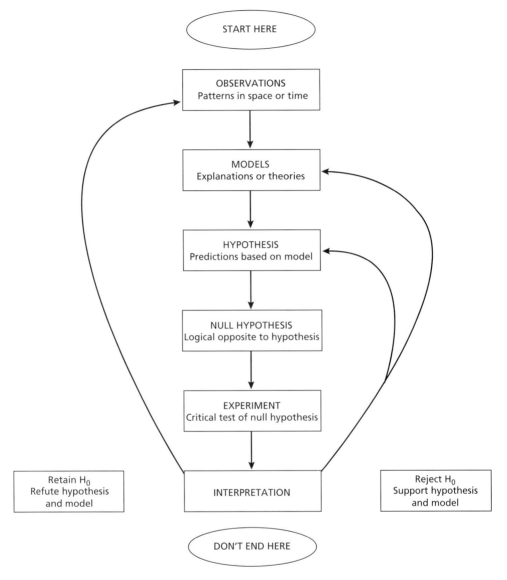

Fig. 1.3 Conceptual model of hypothesis development, including experimentation, interpretation and feedback to support formulation of new testable hypotheses. (From Underwood 1997, by permission of Cambridge University Press.)

predictions arise when we consider three levels simultaneously: (i) the level in question (e.g. community); (ii) the level below which accounts for the mechanisms (e.g. populations interact to generate community-level behaviours); and (iii) the level above that gives context, role, or significance (e.g. the ecosystem, which controls or constrains community-level behaviour).

1.2.3 Multiple working hypotheses

Conceptual models, whether implicit or explicit, serve as the basis for hypothesis formulation. Most scientists gain considerable experience in formulating and testing hypotheses during the course of their careers. Often, this experience is gained during the analysis of specific data sets to test single predefined hypotheses. Adherence to the process of formulating and testing single hypotheses may, however, limit the robustness of ecological research. In a paper that was originally published in 1897 in *The Journal of Geology* (Chamberlin 1897), and was recently (1995) reprinted in the same journal and elsewhere (Hilborn & Mangel 1997), Chamberlin contends that 'in following a single hypothesis the mind is biased by the presumptions of its method toward a single explanatory conception'. He postulates that adequate explanations for complicated phenomenon are often necessarily complex, involving the interaction of several constantly varying elements. Chamberlin (1995) proposes that scientists adopt the method of multiple working hypotheses whereby 'the effort is to bring up into view every rational explanation of the phenomenon in hand and to develop every tenable hypothesis relative to its nature, cause or origin, and to give all of these as impartially as possible a working form and a due place in the investigation'. Although these ideas are more than a century old, they are as worthy of consideration today by ecologists as when originally published.

1.3 Research approaches

Many different types of research approaches have been adopted or, in some cases, pioneered by ecologists. Some approaches are better suited to particular classes of problems than others. Some of the more commonly encountered, although not mutually exclusive, research approaches include focused experimentation; long-term studies, including retrospective studies; large-scale comparative studies; space-for-time substitution; simulation modeling; and ecological restoration (or creation).

1.3.1 Experimentation

Ecology has a rich history in employing innovative laboratory and field experiments. In fact, the field experiments and analyses designed by R.A. Fisher as part of the long-term agricultural ecology programme at Rothamsted Experimental Station in the UK serve as the foundation for analysis of variance and experimental design (Barnett 1994). Well-designed and focused laboratory and field experiments have served as the basis for many fundamental conceptual advances in ecology, leading to the identification of underlying mechanisms that are responsible for population, community and ecosystem structure and function.

At their simplest, ecological experiments might consist of comparing the response of one or more experimental units to a given treatment with one or more control units. However, the range of ecological experiments is incredibly diverse, covering a spectrum of sizes, types of treatments, numbers of replicates and statistical designs. For instance, considering size alone, most ecological experiments range from those performed in laboratory test tubes to 1 m^2 plots to larger field plots and mesocosms. Some notable, and very large, exceptions to this generalization are the acidification and eutrophication experiments performed at the whole-lake scale by Schindler and colleagues in Canada (Schindler *et al.* 1985, 1987), and the diverse, long-term studies conducted by Bormann and Likens at the experimentally deforested catchment at Hubbard Brook Experimental Forest in the northeastern USA (Likens *et al.* 1977; Bormann & Likens 1979). In addition to size differences among experiments, treatments can run the gamut from the conventional application of a single treatment to iterative applications of multiple treatments, such as might commonly be encountered in adaptive management and ecological restoration studies.

Issues related to replication, particularly decisions about the number and locations of replicates, are critical to the success of ecological experiments. In an often cited paper, Hurlbert (1984) reviewed a diverse array of ecological studies and found that many were affected by 'pseudoreplication'—that is, they were either not replicated or the replication was inadequate. Inadequate replication is typically related to a lack of independence in time or space, or to pre-manipulation differences in control and experimental units that persist through the study. In such cases, experimental results are considered 'confounded'. The solution to pseudoreplication is to increase the number of independent replicates that are assigned to each treatment. In reality, though, identification of suitable numbers of appropriately sized and independent replicates is often not a trivial exercise, requiring considerable forethought and field effort.

A variety of experimental designs that have been employed in ecology and other scientific disciplines is illustrated in Fig. 1.4. Comprehensive treatments of experimental design and analysis for ecology are given in texts by Green (1979), Hairston (1991), Manly (1992), Mooney *et al.* (1991), Scheiner and Gurevitch (1993) and Resetarits and Bernardo (1998). Other excellent general references include Kempthorne (1983), Mead (1988), Winer *et al.* (1991) and Sokal and Rohlf (1995). These references contain the background necessary for designing robust ecological experiments. Some non-traditional project designs will be considered in Chapter 7.

Experimentation can be a very powerful approach for distinguishing

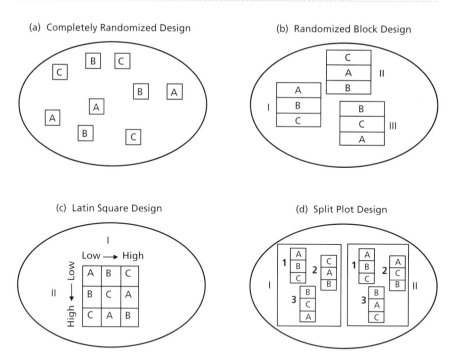

Fig. 1.4 A range of experimental designs that are employed in research studies (A, B, C represent treatments; I, II, III are blocks; and 1,2,3 are plots). (a) Completely random design entails random allocation of three treatments. (b) Randomized complete block design requires that the three treatments be randomly assigned within each of three blocks (i.e. sets of experimental units that are more uniform than randomly chosen experimental units). (c) Latin square design works best in 'controlled' environments where blocking occurs in two directions and all treatments occur in each column and each row. (d) Split plot design is for two or more factors and requires that each level of factor 1 (in this case, 1, 2, 3) is assigned to a whole plot and a second factor (e.g. treatments A, B, C) are randomly allocated to subplots.

causation from correlation. However, Tilman (1989) cautions that there are four primary conceptual problems that are often not considered during the design and interpretation of ecological experiments (particularly those performed in the field) including: (i) transient dynamics (the complex dynamics experienced by a system during its transition from original to experimentally altered states); (ii) indirect effects; (iii) environmental variability, especially through time; and (iv) multiple stable equilibria and site history. These conceptual problems, plus a paucity of long-term experimentation in ecology, may lead to misinterpretation of experimental results, and limit our ability to understand general patterns in nature and to predict natural ecosystem responses to disturbances (Tilman 1989). Consequently, it will now be useful to examine other research approaches that can be employed instead of, or in conjunction with, more conventional ecological experiments.

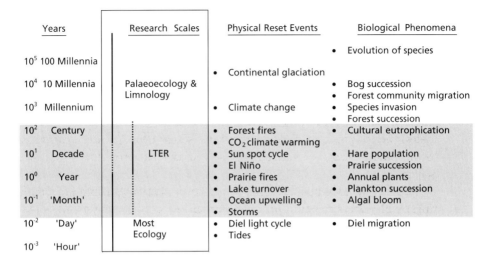

Fig. 1.5 Time scales appropriate for documenting various physical events and biological phenomena. (From *BioScience* 40 (July/August 1990), page 497 (© 1990 American Institute of Biological Sciences) by permission of John Magnuson.)

1.3.2 Long-term studies

Long-term studies are uniquely suited for many questions addressed by ecologists, especially those related to slow processes, rare events or episodic phenomena, processes with high variability, or subtle processes and complex phenomena (Strayer *et al.* 1986; Franklin 1989; Likens 1989; Franklin *et al.* 1990). Effects of broad-scale manipulations, such as the clear-cutting of an entire watershed at Hubbard Brook, can often only be understood through long-term study (Likens 1985). Magnuson (1990) provides an especially striking example from a 132-year record of the duration of ice cover on Lake Mendota, Wisconsin. In a series of graphs with different temporal windows, he illustrates that a single year's observation is relatively uninteresting and provides no insight into system behaviour, whereas progressively longer windows uncover 'hidden' processes like El Niño events, little ice-ages and climate warming trends. The relationship between time scale and the ability to understand ecological structure and function that are 'hidden in the invisible present' is summarized in Fig. 1.5. Despite their obvious importance, long-term studies are, unfortunately, relatively rare in the ecological sciences.

There are several reasons for the paucity of long-term studies. Statistical differentiation of trends or responses from 'noise' may require an extremely long period of record (e.g. years to decades and longer for slow processes and long-lived species) and a concomitant long-term commitment of personnel and funds which may preclude their widespread application. Long-term study design requires careful consideration of issues related to data and sampling consistency, quality assurance and data management, and tradeoffs among temporal, spatial and thematic (i.e. numbers and types of parameters) resolution. Despite these drawbacks, there is often no substitution for long-term study. In cases where new long-term studies cannot be initiated, it may, nevertheless, be insightful to compare responses observed following 'treatments' to the natural range of variability observed in long-term studies (e.g. see Schindler 1991).

Retrospective studies, including dendrochronological (e.g. analyses of tree rings) and palaeoecological (e.g. analyses of ice cores, sediment cores) analyses, represent a special category of long-term studies (Davis 1989) that may be especially useful for reconstructing site histories and identifying natural ranges in variability. For instance, dendrochronological techniques have been used to examine past climate (e.g. Fritts 1976; Briffa *et al.* 1990; Graumlich 1993), date environmental processes such as forest disturbances (e.g. Banks 1991; Fritts & Swetnam 1989), monitor long-term trends in air pollution (e.g. Cogbill 1977; Sutherland & Martin 1990), detect increases in CO_2 (e.g. LaMarche *et al.* 1984; Jacoby & D'Arrigo 1997), examine multi-century patterns in insect damage (e.g. Swetnam & Lynch 1993; Weber 1997), characterize long-term stand dynamics (e.g. Foster 1988; Payette & Gagnon 1979), support multi-century reconstruction of forest fire dynamics (e.g. Engelmark 1984; Swetnam 1993) and date various earth surface processes like volcanic eruptions (e.g. Baillie & Munro 1988) and glacier movements (e.g. Luckman 1994).

1.3.3 Comparative studies

New insights and understanding, as well as improved predictive power for ecology, can be gained through the design and implementation of large-scale comparative studies. Comparative studies might consider ecological responses of populations, communities or ecosystems to similar treatments 'replicated' across gradients, landscapes or larger areas (Pace 1993). Comparative studies may be either planned experiments or after-the-fact syntheses of existing studies. Pace (1993), as an example of the latter, cites a study by Peierls *et al.* (1991) in which they evaluated nitrate concentrations in 42 major rivers of the world.

Although large-scale comparative studies can be used for the identification of general patterns, they may not provide the mechanistic understanding associated with more focused experiments. Conversely, mechanistic studies often neglect unexpected ecological interactions that emerge as being important in comparative studies (Pace 1993). Strengths and weaknesses of the comparative study approach are discussed more fully by Carpenter *et al.* (1991) and Pace (1993).

1.3.4 Space-for-time substitution

In ecosystems with long-lived species, succession and the development of other ecological patterns and processes may occur slowly, over periods lasting from decades to centuries. Study of such changes may only be practical where it can be demonstrated that space and time are surrogates for one another. For example, studies examining primary and secondary succession, recovery from disturbance, and palaeolimnology are often based on the premise that spatial sequences are homologous with temporal sequences and that temporal trends can be extrapolated from chronosequences (sites or samples of different ages) obtained by sampling populations, communities, and ecosystems of different ages (e.g. Charles & Smol 1988; Pickett 1989). Like comparative studies, space-for-time substitution studies may be most effective for identifying general trends and generating hypotheses, but may not expose underlying mechanisms (e.g. see Lauenroth & Sala 1992 for discussion of weaknesses related to examining above ground net primary production). A comprehensive review of the strengths and weaknesses of space-for-time substitution is provided by Pickett (1989).

1.3.5 Simulation modelling studies

Simulation models are tools of varying levels of complexity that are used for characterizing and understanding ecological patterns and processes (Gillman & Hails 1997). Models typically consist of one or more related mathematical equations that are designed to encapsulate the salient features of real-world processes. Models vary considerably in their degree of sophistication; population growth models, for instance, may be purely deterministic or may also incorporate elements of environmental stochasticity. As computers have become more powerful, simulation models have become increasingly complex. Models are now being linked with geographic information systems to examine processes like groundwater flows and primary production at scales ranging from the ecosystem to region and beyond (Steyaert & Goodchild 1994). Significant effort has also been devoted recently to developing population, community and ecosystem models that are based on the properties of individuals, including their behaviour and interactions with others (DeAngelis & Gross 1992).

Simulation modelling has several uses in ecological research:

1 a good model represents the culmination of efforts to synthesize what we know about a system or process, thereby serving as a good thought-organizing device;

2 models can be experimented with by modifying model components and recording the effects on the rest of the system; and

3 good models allow us to forecast how dynamic systems will change in the future (Hannon & Ruth 1997).

Simulation modelling may be especially appropriate for assessing potential effects of large, expensive manipulations prior to their implementation, and examining potential ecological outcomes under different climate and disturbance regimes (e.g. Costanza *et al.* 1990). In such cases, models are invaluable in that they allow scientists to alter virtual populations and ecosystems with no real adverse ecological effects. Furthermore, they allow periods ranging from days to centuries to be compressed into the much shorter time required for model execution. Although simulations may be run independently as part of a stand-alone research project, model validation may require access to high quality data collected using one or more of the other research approaches described in this chapter (also see Kirchner 1994).

1.3.6 Ecological restoration

Ecological restoration can be a very powerful test of basic ecological theory (e.g. Bradshaw 1987; Harper 1987; Jordan *et al.* 1987). Harper (1987), for example, suggested that ecology can benefit from attempts by restoration practitioners to design and create new communities and simultaneously test ideas or answer fundamental ecological questions. Some examples of relevant research topics include determining the relationship between species diversity and community stability and resilience; understanding the roles that animals and mutualists play in succession; and documenting the effects of age structure and genetic diversity of component species on community properties (see Bell *et al.* 1997; Ehrenfeld & Toth 1997; Montalvo *et al.* 1997; Palmer *et al.* 1997 for landscape, ecosystem, population and community research perspectives, respectively).

Despite the attractiveness of viewing ecological restoration projects as 'experiments', the lack of control in designing and implementing an experiment can present an enormous challenge. Many restoration projects are mandated for specific reasons and must proceed along specific timetables, regardless of whether or not scientists are involved. Consequently, ecologists may become involved in a specific project long after it has been designed or, in some cases, long after the project has been completed. Project objectives, proposed implementation plans, time frames and site(s) are often established prior to on-the-ground work and all have direct bearing on research planning.

Other project parameters may evolve during the course of the project. For example, 'treatment' frequency (e.g. single, repeated, multiple), intensity (spot herbicide application, broadcast spraying), and type (e.g. herbicide application, fire, cutting and chopping) may vary throughout the project depending upon perceived effectiveness. Despite experimental design challenges, efforts to create or restore ecological systems represent the ultimate test of our understanding of ecological structure and function. Parmenter and MacMahon (1983), for instance, describe a carefully designed mine restoration experiment whereby different planting patterns resulted in different spatial and age-structure vegetation patterns and concomitant changes in biota and soil characteristics.

Perceived 'failures' of restoration experiments are less likely to be documented in the literature than successes. Nevertheless, reporting the results of experiments that are viewed as less than successful should minimally lead to new hypotheses and models, and could lead to new and more robust ecological paradigms. Middleton (1999), for instance, provides examples of wetland restoration projects that did not meet their objectives because of a failure to adequately incorporate natural disturbance regimes like flood pulsing. Zedler (1996) uses specific case studies to document problems with wetland mitigation efforts in southern California, recommends regional wetland restoration planning, and highlights the need for post-mitigation monitoring to track successes and failures.

1.4 Research design and implementation issues

1.4.1 Specifying the research domain

Often the argument is made that comprehensive understanding of ecological phenomena requires **long-term** monitoring of **salient patterns and processes** in adequately **replicated** control and experimental units at **appropriate spatial and temporal scales** using sound sampling design and statistical analyses. Although this line of reasoning is based on fundamental ecological principles and the existing knowledge base, this theoretical optimum in research design can rarely be achieved. With respect to ecological structure and function, 'long-term' may imply decades to centuries, a period of study that generally is not feasible due to a variety of political and funding constraints. Consequently, ecological understanding often emerges from the synthesis of results from short-term studies that, individually, represent only snapshots of the long-term system dynamics.

Ecologists cannot measure everything and, often, cannot define 'salient patterns and processes' in many ecosystems (Mooney 1991; Ray *et al.* 1992).

Many 'treatments' are not suitable for replication since candidate sites for replication may not be readily apparent, accessible or exist. For instance, in the Hubbard Brook example presented earlier, the watershed that was clear-cut was not replicated. Furthermore, Likens (1985) highlighted the fact that the 'control' was not really a control in the experimental sense, but rather represented a reference or benchmark to be used for comparison purposes.

How then does one establish the research domain? In other words, how do we choose appropriate spatial and temporal scales and narrow down the list of parameters to measure (i.e. thematic scale)? Generally, there are no straightforward answers to these questions. Allen and Starr (1982) have cautioned that adopting a particular point of view towards the environment affects the research design, results, and interpretation. However, in establishing the research domain for a project, this is exactly what we are doing.

The process of defining the research domain generally entails several activities. First, scientists develop a comprehensive knowledge of the subject area and scour the literature to ascertain what has been done in the past. Consciously or subconsciously we often record what has been tried and proven successful in the past. Second, based on past studies, knowledge of the ecosystem (or study area) and a certain amount of intuition, scientists speculate on the minimum and optimum spatial, temporal and thematic scales of resolution. In the spatial realm, we are interested in the size (area extent), number, resolution, spatial configuration and the location of experimental units, and control or reference sites. Temporally, we are interested in specifying minimum and optimal frequency of sampling (e.g. hourly, daily, monthly) and duration of the study, taking into account practical and financial constraints. Determining the thematic resolution can be the most difficult since it can entail consideration of multiple types and intensities of treatment levels, as well as causal and constraining factors, and the various multi-way interactions. Finally, we must consider the various possibilities in concert. Generally, we weigh the tradeoffs of different scales of resolution, incorporating potential costs in money, personnel and other pragmatic concerns where known. Often, many of the choices related to the research domain are constrained in advance. For instance, funding levels, timing and limitations on length of grant support can all constrain research design options. In addition, the lack of particular types of skills, inaccessibility of certain instruments and other limitations figure (at least subconsciously) into our specification of the research domain. Green (1979) provides a cogent discussion of many other decisions that must be made in targeting the research domain, especially variable selection.

Specifying the research domain is usually the most challenging and important task performed as part of a research project. Often, the selection of an appropriate sampling design is relatively straightforward, once the research domain has been adequately specified. However, at least three other considerations can have direct bearing on the final design of a research project. These include statistical power, standard methods and pilot studies.

1.4.2 Statistical power

Statistical power represents the ability of an experiment to detect treatment effects of a given magnitude (i.e. correctly reject the null hypothesis). Evaluation of statistical power, which can be critical to the selection of a final experimental design, requires specification of the experiment-wise level of significance (e.g. $p = 0.05$), the sample size, magnitude of the treatment effect one wishes to be able to detect and the within-class variance associated with the observations (Winer *et al.* 1991). Statistical power is frequently evaluated in the context of determining the sample size required to detect an effect of a given magnitude. For more information on determining statistical power, ascertaining sample sizes and related topics see Winer *et al.* (1991) and Underwood (1997). Various software programs used in calculating statistical power have been reviewed by Thomas and Krebs (1997).

1.4.3 Standardization

Because temporal and spatial variability are inherent in ecological patterns and processes, our ecological understanding increases when we replicate experiments (or observations) in different habitats and at different times. Analysis and interpretation of ecological studies are often facilitated when standard methods are employed. For instance, different results are frequently obtained when we use different field and laboratory methods to measure identical phenomenon. Unless a particular method is inter-calibrated with other methods, there may be no way of relating data obtained in one study to others. For these reasons, most studies of a particular type rely upon a limited number of methods, sampling gear (e.g. mesh size of plankton nets), and instrumentation. When 'standard' methods exist and have been well documented in the literature, ecologists can benefit in numerous ways from adopting these methods; for example, comparison with other studies is facilitated, documentation of the methods for publications and metadata (see Chapter 5) is easier, and costs are often lower.

Some references for standard methods commonly employed in ecological research are included in Table 1.1. This list is by no means exhaustive, nor does it obviate the need to critically evaluate appropriate methods for a particular study on a case-by-case basis. In some cases, standard methods may not exist or they may be inappropriate. Furthermore, it should be noted that there

Table 1.1 Examples of references to standard methods commonly employed in ecological research.

Data type	Examples	Reference
Soil and below-ground systems	sediment cores (e.g. paleoclimatology and palaeoceanography)	Kemp (1996)
	soil ecology and soil sampling/analysis	Klute (1986); Carter(1993); Weaver (1994); Schinner *et al.* (1996); Robertson *et al.* (1999)
	root systems	Boehm (1987)
Animal population ecology and biodiversity and censusing	birds	Bibby *et al.* (1992); Pyle (1997)
	mammals	Wilson *et al.* (1996)
	amphibians	Heyer *et al.* (1993)
	radiotelemetry and mark/recapture	Seber (1973); White & Garrott (1990); Skalski & Robson (1992); Keeger (1996)
	terrestrial vertebrates	Davis (1983)
	wildlife and habitats	Bookhout (1994)
Plant ecology	plant ecology	Pandeya (1968); Bonham (1989); Kent & Coker (1994)
	dendrochronology	Fritts & Swetnam (1989); Cook & Kairiukstis (1990)
	plant physiology	Pearcy *et al.* (1989); Lassoie & Hinckley (1991)
Chemical	chemical ecology	Millar & Haynes (1999)
	ecological toxicology	Hammons (1981); Hoffman *et al.* (1995)
Water and air	aquatic microbial ecology	Kemp *et al.* (1993)
	water quality	Greenberg (1995)
	stream ecology	Hauer & Lamberti (1996)
	hydrology	Maidment (1993)
	air pollution, assessment and modelling	Weber (1982)
Landscape ecology	landscape heterogeneity	Turner & Gardner (1996)
	digital image processing	Jensen (1996)
	geographic information system analyses	Burrough & McDonnell (1998); Johnston (1998)
Miscellaneous	general ecology	Bower (1997); Krebs (1998)
	diversity indices	Magurran (1988)
	integrated monitoring	Sykes & Lane (1996); IM Programme Centre (1998)

is always a need for new and improved (e.g. more inexpensive, faster, more accurate) methods. Consequently, the recommendations in this book concerning the development and adoption of standards should be considered with these caveats in mind.

1.4.4 Pilot studies

Prior to formally specifying the research design (methods, parameters, replication etc.), conducting a pilot study can often be beneficial. Objectives of a pilot study can be severalfold. First, pilot studies can be conducted to test that equipment and sampling gear perform as expected (i.e. a 'shakedown cruise'). Second, pilot studies can be useful for planning coordination among various parties involved in field campaigns. Third, and possibly the most useful objective from a research design standpoint, is to ascertain within-class variance for some or all of the parameters to be measured (see section 1.4.2). Although treatments, if any, may not have been applied, some indication of the intrinsic variability can be extremely useful for designing effective field and laboratory studies. In lieu of variability estimates from pilot studies or the literature, one must arbitrarily decide upon the number of replicates to be taken. If one knows a priori that the variability of a critical parameter is substantially higher or lower than others, then the effort spent replicating observations can be adjusted accordingly.

1.5 Next steps

This chapter has examined several different approaches to the design of ecological studies. It is now time to turn our attention to several data-related issues that arise during project implementation. Consequently, the next five chapters provide comprehensive treatments of data management principles (Chapter 2), scientific databases (Chapter 3), data quality assurance (Chapter 4), data documentation (Chapter 5) and archiving ecological data and information (Chapter 6). Finally, in Chapters 7 and 8 we revisit some of the ideas introduced in this chapter, including a discussion of the various approaches that can be used to transform data into information and, ultimately, knowledge. It is hoped that the information presented in this book will contribute to a new paradigm for how we conduct ecological research, one that views data as a resource and promotes stewardship, recycling and sharing of data.

1.6 References

Allen, T.F.H. & Hoekstra, T.W. (1992) *Toward a Unified Ecology*. Columbia University Press, New York.

Allen, T.F.H. & Starr, T.B. (1982) *Hierarchy: Perspectives for Ecological Complexity*. University of Chicago Press, Chicago, IL.

Baillie, M.G.L. & Munro, M.A.R. (1988) Irish tree rings, Santorini and volcanic dust veils. *Nature* **332**, 344–346.

Banks, J.C.G. (1991) A review of the use of tree rings for the quantification of forest disturbances. *Dendrochronologia* **9**, 51–70.

Barnett, V. (1994) Statistics and the long-term experiments: past achievements and future challenges. In: *Long-term Experiments in Agricultural and Ecological Studies*. (eds R.A. Leigh & A.E. Johnston), pp. 165–183. CAB International, Wallingford, Oxon, UK.
Bell, S.S., Fonseca, M.S. & Motten, L.B. (1997) Linking restoration and landscape ecology. *Restoration Ecology* **5**, 318–323.
Bernstein, B.B. & Goldfarb, L. (1995) A conceptual tool for generating and evaluating ecological hypotheses. *BioScience* **45**, 32–39.
Bibby, C.J., Burgess, N.D. & Hill, D.A. (1992) *Bird Census Techniques*. Academic Press, London.
Boehm, W. (1987) *Methods of Studying Root Systems*. Springer-Verlag, New York.
Bonham, C.D. (1989) *Measurements for Terrestrial Vegetation*. John Wiley & Sons, New York.
Bookhout, T.A. (1994) *Research and Management Techniques for Wildlife and Habitats*. The Wildlife Society, Bethesda, MD.
Bormann, F.H. & Likens, G.E. (1979) *Pattern and Process in a Forested Ecosystem*. Springer-Verlag, New York.
Bower, J.E. (1997) *Field and Laboratory Methods for General Ecology*. 4th edn. McGraw-Hill, New York.
Bradshaw, A.D. (1987) Restoration: an acid test for ecology. In: *Restoration Ecology: A Synthetic Approach to Ecological Research*. (eds W.R. Jordan III, M.E. Gilpin & J.D. Aber), pp. 23–29. Cambridge University Press, Cambridge, UK.
Briffa, K.R., Bartholin, T.S., Eckstein, D. *et al.* (1990) A 1400-year tree-ring record of summer temperatures in Fennoscandia. *Nature* **346**, 434–439.
Brown, J.H. & Roughgarden, J. (1990) Ecology for a changing earth. *Bulletin of the Ecological Society of America* **71**, 173–188.
Burrough, P.A. & McDonnell, R.A. (1998) *Principles of Geographic Information Systems*. Oxford University Press, Oxford, UK.
Carpenter, S.R., Frost, T.M., Kitchell, J.F. *et al.* (1991) Patterns of primary production and herbivory in 25 North American lake ecosystems. In: *Comparative Analysis of Ecosystems* (eds J. Cole, G. Lovett & S. Findlay), pp. 67–96. Springer-Verlag, New York.
Carter, M.R. (1993) *Soil Sampling and Methods of Analysis*. Lewis Publishers, Boca Raton, FL.
Chamberlin, T.C. (1897) The method of multiple working hypotheses. *The Journal of Geology* **5**, 837–848.
Chamberlin, T.C. (1995) The method of multiple working hypotheses. *The Journal of Geology* **103**, 349–354.
Charles, D.F. & Smol, J.P. (1988) New methods for using diatoms and chrysophytes to infer past pH of low-alkalinity lakes. *Limnology and Oceanography* **33**, 1451–1462.
Cogbill, C.V. (1977) The effect of acid precipitation on tree growth in eastern North America. *Water, Air, and Soil Pollution* **8**, 89–93.
Colwell, R.K. (1995) Ecological Society of America special committee on ESA communications in the electronic age. *Bulletin of the Ecological Society of America* **76**, 120–131.
Cook, E.R. & Kairiukstis, L.A. (eds) (1990) *Methods of Dendrochronology—Applications in the Environmental Sciences*. Kluwer Academic Publishers and International Institute for Applied Systems Analysis, Dordrecht, The Netherlands.
Costanza, R., Sklar, F.H. & White, M.L. (1990) Modeling coastal landscape dynamics. *BioScience* **40**, 91–107.
Davis, D.E. (1983) *CRC Handbook of Census Methods for Terrestrial Vertebrates*. CRC Press, Boca Raton, FL.
Davis, M.B. (1989) Retrospective studies. In: *Long-term Studies in Ecology: Approaches and Alternatives* (ed G.E. Likens), pp. 71–89. Springer-Verlag, New York.
DeAngelis, D.L. & Gross, L.J. (eds) (1992) *Individual-based Models and Approaches in Ecology*. Chapman & Hall, New York.
Ehrenfeld, J.G. & Toth, L.A. (1997) Restoration ecology and the ecosystem perspective. *Restoration Ecology* **5**, 307–317.
Engelmark, O. (1984) Forest fires in the Maddus National Park (northern Sweden) during the past 600 years. *Canadian Journal of Botany* **62**, 893–898.

Forman, R.T. (1995) *Land Mosaics: The Ecology of Landscapes and Regions*. Cambridge University Press, Cambridge, UK.

Foster, D.R. (1988) Disturbance history, community organization and vegetation dynamics of the old-growth Pisgah Forest, south-western New Hampshire, USA. *Journal of Ecology* **76**, 105–134.

Franklin, J.F. (1989) Importance and justification of long-term studies in ecology. In: *Long-term Studies in Ecology: Approaches and Alternatives* (ed G.E. Likens), pp. 3–19. Springer-Verlag, New York.

Franklin, J.F., Bledsoe, C.S. & Callahan, J.T. (1990) Contributions of the long-term ecological research program. *BioScience* **40**, 509–523.

Fritts, H.C. (1976) *Tree Rings and Climate*. Academic Press, New York.

Fritts, H.C. & Swetnam, T.W. (1989) Dendroecology: a tool for evaluating variations in past and present forest environments. *Advances in Ecological Research* **19**, 111–188.

Gillman, M. & Hails, R. (1997) *An Introduction to Ecological Modelling: Putting Practice into Theory*. Blackwell Science, Oxford, UK.

Gosz, J.R. (1994) Sustainable Biosphere Initiative: data management challenges. In: *Environmental Information Management and Analysis: Ecosystem to Global Scales* (eds W.K. Michener, J.W. Brunt & S.G. Stafford), pp. 27–39. Taylor & Francis, Ltd., London, UK.

Graumlich, L.J. (1993) A 1000-year record of temperature and precipitation in the Sierra Nevada. *Quaternary Research* **39**, 249–255.

Green, R.H. (1979) *Sampling Design and Statistical Methods for Environmental Biologists*. John Wiley & Sons, Inc., New York.

Greenberg, A.E. (1995) *Standard Methods for the Examination of Water and Wastewater*. (19th edn). American Public Health Association, Washington.

Gross, K.L., Pake, C.E., Allen, E. et al. (1995a) *Final report of the Ecological Society of America Committee on the Future of Long-term Ecological Data (FLED), Volume I: Text of the report*. Ecological Society of America, Washington, DC. (http://www.sdsc.edu/~ESA/FLED/FLED.html).

Gross, K.L., Pake, C.E., O'Neill, A. et al. (1995b) *Final report of the Ecological Society of America Committee on the Future of Long-term Ecological Data (FLED), Volume II: Directories to Sources of Long-term Ecological Data*. Ecological Society of America, Washington, DC. (http://www.sdsc.edu/~ESA/FLED/FLED.html/rep_vol2.html).

Hairston, N.G. (1991) *Ecological Experiments: Purpose, Design, and Execution*. Cambridge University Press, Cambridge, UK.

Hammons, A.S. (1981) *Methods for Ecological Toxicology*. Technomic Publishing Company, Lancaster, PA.

Hannon, B. & Ruth, M. (1997) *Modeling Dynamic Systems*. Springer-Verlag, New York.

Harper, J.L. (1987) *Restoration Ecology: A Synthetic Approach to Ecological Research* (eds W.R. Jordan, III, M.E. Gilpin & J.D. Aber), pp. 35–45. Cambridge University Press, Cambridge, UK.

Hauer, F.R. & Lamberti, G.A. (1996) *Methods in Stream Ecology*. Academic Press, San Diego, CA.

Heyer, W.R., Donnelly, M.A., McDiarmid, R.W., Hayek, L.C. & Foster, M.S. (1993) *Measuring and Monitoring Biological Diversity: Standard Methods for Amphibians*. Smithsonian Institute Press, Washington, DC.

Hilborn, R. & Mangel, M. (1997) *The Ecological Detective: Confronting Models with Data*. Princeton University Press, Princeton, NJ.

Hoffman, D.J., Rattner, B.A., Burton Jr., G.A. & Cairns, J., Jr. (1995) *Handbook of Ecotoxicology*. CRC Press, Boca Raton, FL.

Hurlbert, S.H. (1984) Pseudoreplication and the design of ecological field experiments. *Ecological Monographs* **54**, 187–211.

Huston, M.A. (1994) *Biological Diversity*. Cambridge University Press, Cambridge, UK.

IM Programme Centre (1998) *Manual for Integrated Monitoring*. ICP IM Programme Centre, Finnish Environment Institute, Helsinki, Finland.

Jacoby, G.C. & D'Arrigo, R.D. (1997) Tree rings, carbon dioxide, and climate change. *Proceedings of the National Academy of Sciences of the United States of America* **94**, 8350–8353.

Jensen, J.R. (1996) *Introductory Digital Image Processing: A Remote Sensing Perspective*. 2nd edn. Prentice Hall, Upper Saddle River, NJ.

Johnston, C.A. (1998) *Geographic Information Systems in Ecology*. Blackwell Science, Oxford, UK.

Jordan, III, W.R., Gilpin, M.E. & Aber, J.D. (1987) Restoration ecology: ecological restoration as a technique for basic research. In: *Restoration Ecology: A Synthetic Approach to Ecological Research* (eds W.R. Jordan III, M.E. Gilpin & J.D. Aber), pp. 3–21. Cambridge University Press, Cambridge, UK.

Kareiva, P. & Anderson, M. (1988) Spatial aspects of species interactions: the wedding of models and experiments. In: *Community Ecology* (ed A. Hastings), pp. 35–50. Springer-Verlag, New York.

Keeger, T.J. (1996) *Handbook of Wildlife Chemical Immobilization*. International Wildlife Veterinary Services, Inc., Laramie, WY.

Kemp, A.E. (1996) *Paleoclimatology and Palaeoceanography from Laminated Sediments*. American Association of Petroleum Geologists, Tulsa, OK.

Kemp, P.F., Sherr, B.F., Sherr, E.B. & Cole, J.J. (1993) *Handbook of Methods in Aquatic Microbial Ecology*. Lewis Publishers, Boca Raton, FL.

Kempthorne, O. (1983) *The Design and Analysis of Experiments*. Robert E. Krieger Publishing Company, Malabar, FL.

Kent, M. & Coker, P. (1994) *Vegetation Description and Analysis: A Practical Approach*. John Wiley and Sons, New York.

Kirchner, T.B. (1994) Data management and simulation modelling. In: *Environmental Information Management and Analysis: Ecosystem to Global Scales* (eds W.K. Michener, J.W. Brunt & S.G. Stafford), pp. 357–375. Taylor & Francis, Ltd., London.

Klute, A. (ed) (1986) *Methods of Soil Analysis: Part 1: Physical and Mineralogical Methods*. 2nd edn. Number 9 in the Series Agronomy. Soil Science Society of America, Inc., Madison, WI.

Krebs, C.J. (1998) *Ecological Methodology*. 2nd edn. Addison-Wesley Publishing Corporation, Reading, MA.

Kuhn, T. (1970) *The Structure of Scientific Revolutions*. 2nd edn. University of Chicago Press, Chicago, IL.

LaMarche, V.C., Jr., Graybill, D.A., Fritts, H.C. & Rose, M.R. (1984) Increasing atmospheric carbon dioxide: tree ring evidence for growth enhancement in natural vegetation. *Science* **225**, 1019–1021.

Lassoie, J.P. & Hinckley, T.M. (eds) (1991) *Techniques and Approaches in Forest Tree Ecophysiology*. CRC Press, Boca Raton, FL.

Lauenroth, W.K. & Sala, O.E. (1992) Long-term forage production of North American short-grass steppe. *Ecological Applications* **2**, 397–403.

Levin, S.A. (1992) The problem of pattern and scale in ecology. *Ecology* **73**, 1943–1967.

Likens, G.E. (1985) An experimental approach for the study of ecosystems. *Journal of Ecology* **73**, 381–396.

Likens, G.E. (ed) (1989) *Long-term Studies in Ecology*. Springer-Verlag, New York.

Likens, G.E., Bormann, F.H., Pierce, R.S., Eaton, J.S. & Johnson, N.M. (1977) *Biogeochemistry of a Forested Ecosystem*. Springer-Verlag, New York.

Luckman, B.H. (1994) Glacier fluctuation and tree-ring records for the last millennium in the Canadian Rockies. *Quaternary Science Reviews* **12**, 441–450.

Maidment, D.R. (1993) *Handbook of Hydrology*, McGraw-Hill, New York.

Magnuson, J.J. (1990) Long-term ecological research and the invisible present. *BioScience* **40**, 495–501.

Magurran, A.E. (1988) *Ecological Diversity and its Measurement*. Princeton University Press, Princeton, NJ.

Manly, B.F.J. (1992) *The Design and Analysis of Research Studies*. Cambridge University Press, Cambridge, UK.
Mead, R. (1988) *The Design of Experiments*. Cambridge University Press, Cambridge, UK.
Middleton, B. (1999) *Wetland Restoration, Flood Pulsing, and Disturbance Dynamics*. Wiley & Sons, Inc., New York.
Millar, J.G. & Haynes, K.F. (1999) *Methods in Chemical Ecology*, Vol. 1 & 2. Chapman & Hall, New York.
Montalvo, A.M., William, S.L., Rice, K.J. et al. (1997) Restoration biology: a population biology perspective. *Restoration Ecology* **5**, 277–290.
Mooney, H. (1991) Biological response to climate change: an agenda for research. *Ecological Applications* **1**, 112–117.
Mooney, H.A., Medina, E., Schindler, D.W., Schulze, E.D. & Walker, B.H. (1991) *Ecosystem Experiments*. John Wiley & Sons Ltd., Chichester, UK.
Morgan, P., Aplet, G.H., Haufler, J.B., Humphries, H.C., Moore, M.M. & Wilson, W.D. (1994) Historical range of variability: a useful tool for evaluating ecosystem change. In: *Assessing Forest Ecosystem Health in the Inland West*. (eds R.N. Sampson & D.J. Adams), pp. 87–111. The Haworth Press, Inc., New York.
National Research Council. (1991) *Solving the Global Change Puzzle: A US Strategy for Managing Data and Information*. National Academy Press, Washington, DC.
National Research Council. (1993) *A Biological Survey for the Nation*. National Academy Press, Washington, DC.
National Research Council. (1995a) *Finding the Forest in the Trees*. National Academy Press, Washington, DC.
National Research Council. (1995b) *Preserving Scientific Data on Our Physical Universe: A New Strategy for Archiving the Nation's Scientific Information Resources*. National Academy Press, Washington, DC.
National Science Foundation. (1994) *Grant Proposal Guide*. National Science Foundation, Arlington, VA.
Pace, M.L. (1993) Forecasting ecological responses to global change: the need for large-scale comparative studies. In: *Biotic Interactions and Global Change*. (eds P.M. Kareiva, J.G. Kingsolver & R.B. Huey), pp. 356–363. Sinauer Associates Inc., Sunderland, MA.
Palmer, M.A., Ambrose, R.F., & Poff, N.L. (1997) Ecological theory and community restoration ecology. *Restoration Ecology* **5**, 291–300.
Pandeya, S.C. (1968) *Research Methods in Plant Ecology*. Asia Publishing House, New York.
Parmenter, R.R. & MacMahon, J.A. (1983) Factors determining the abundance and distribution of rodents in a shrub-steppe ecosystem: the role of shrubs. *Oecologia* **59**, 145–156.
Payette, S. & Gagnon, R. (1979) Tree-line dynamics in Ungava peninsula, northern Quebec. *Holarctic Ecology* **2**, 239–248.
Pearcy, R.W., Ehleringer, J.R., Mooney, H.A. & Rundel, P.W. (1989) *Plant Physiological Ecology: Field Methods and Instrumentation*. Chapman & Hall, London.
Peierls, B.L., Caraco, N.F., Pace, M.L. & Cole, J.J. (1991) Human influence on river nitrogen. *Nature* **350**, 386–387.
Pickett, S.T.A. (1989) Space-for-time substitution as an alternative to long-term studies. In: *Long-term Studies in Ecology: Approaches and Alternatives*. (ed. G.E. Likens), pp. 110–135. Springer-Verlag, New York.
Pickett, S.T.A., Parker, V.T. & Fiedler, P. (1992) The new paradigm in ecology: implications for conservation biology above the species level. In: *Conservation Biology: The Theory and Practice of Nature Conservation, Preservation, and Management*. (eds P. Fiedler & S. Jain), pp. 65–88. Chapman and Hall, New York.
Pickett, S.T.A., Kolasa, J. & Jones, C.G. (1994) *Ecological Understanding*. Academic Press, San Diego, CA.
Pickett, S.T.A. & Ostfield, R.S. (1995) The shifting paradigm in ecology. In: *A New Century for Natural Resources Management*. (eds R.L. Knight & S.F. Bates), pp. 261–278. Island Press, Washington, DC.

Pimm, S.L. (1991) *The Balance of Nature? Ecological Issues in the Conservation of Species and Communities*. University of Chicago Press, Chicago, IL.

Popper, K.R. (1968) *The Logic of Scientific Discovery*. Hutchinson, London.

Pyle, P. (1997) *Identification Guide to North American Birds*. Slate Creek Press, Bolinas, CA.

Ray, G.C., Hayden, B.P., Bulger, Jr., A.J. & McCormick-Ray, M.G. (1992) Effects of global warming on the biodiversity of coastal-marine zones. In: *Global Warming and Biological Diversity*. (eds R.L. Peters & T.E. Lovejoy), pp. 91–104. Yale University Press, New Haven, CT.

Resetarits, Jr., W.J. & Bernardo, J. (eds) (1998) *Experimental Ecology*. Oxford University Press, New York.

Robertson, G.P., Coleman, D.C., Bledsoe, C.S. & Sollins, P. (1999) *Standard Soil Methods for Long-Term Ecological Research*. Oxford University Press, New York.

Scheiner, S.M. & Gurevitch, J. (eds) (1993) *Design and Analysis of Ecological Experiments*. Chapman & Hall, New York.

Schindler, D.W. (1991) Whole-lake experiments at the Experimental Lakes Area. In: *Ecosystem Experiments SCOPE 45*. (eds H.A. Mooney, E. Medina, D.W. Schindler, E. Shulze & B.J. Walker), pp. 121–139. John Wiley & Sons, Inc., Chichester, UK.

Schindler, D.W., Mills, K.H., Malley, D.F. *et al.* (1985) Long-term ecosystem stress: the effects of years of acidification on a small lake. *Science* **228**, 1395–1401.

Schindler, D.W., Hesslein, R.H. & Turner, M.A. (1987) Exchange of nutrients between sediments and water after 15 years of experimental eutrophication. *Canadian Journal of Fisheries and Aquatic Sciences* **44**, 26–33.

Schinner, F., Kandeler, E., Margesin, R., & Ohliner, R. (1996) *Methods in Soil Biology*. Springer-Verlag, New York.

Seber, G.A.F. (1973) *Estimation of Animal Abundance*. Hafner, New York.

Skalski, J.R. & Robson, D.S. (1992) *Techniques for Wildlife Investigations: Design and Analysis of Capture Data*. Academic Press.

Sokal, R.R. & Rohlf, F.J. (1995) *Biometry*. 3rd edn. W.H. Freeman & Company, New York.

Steyaert, L.T. & Goodchild, M.F. (1994) Sustainable Biosphere Initiative: data management challenges. In: *Environmental Information Management and Analysis: Ecosystem to Global Scales*. (eds W.K. Michener, J.W. Brunt & S.G. Stafford), pp. 333–355. Taylor & Francis, Ltd., London.

Strayer, D.S., Glitzenstein, J.S., Jones, C.G. *et al.* (1986) *Long-term Ecological Studies: An Illustrated Account of Their Design, Operation, and Importance to Ecology*. Institute of Ecosystem Studies, Millbrook, NY.

Sutherland, E.K. & Martin, B. (1990) Growth response of *Pseudotsuga menziesii* to air pollution from copper smelting. *Canadian Journal of Forest Research* **20**, 1020–1030.

Swetnam, T.W. (1993) Fire history and climate change in giant sequoia groves. *Science* **262**, 885–889.

Swetnam, T.W. & Lynch, A.M. (1993) Multicentury, regional-scale patterns of western spruce budworm outbreaks. *Ecological Monographs* **63**, 399–424.

Sykes, J.M. & Lane, A.M.J. (eds) (1996) *The UK Environmental Change Network: Protocols for Standard Measurements at Terrestrial Sites*. Stationery Office, London.

Thomas, L. & Krebs, C.J. (1997) A review of statistical power analysis software. *Bulletin of the Ecological Society of America* **78**, 126–139.

Tilman, D. (1989) Ecological experimentation: strengths and conceptual problems. In: *Long-term Studies in Ecology: Approaches and Alternatives*. (ed G.E. Likens), pp. 136–157. Springer-Verlag, New York.

Turner, M.G. & Gardner, R.H. (1991) *Quantitative Methods in Landscape Ecology*. Springer-Verlag, New York.

Underwood, A.J. (1997) *Experiments in Ecology: Their Logical Design and Interpretation Using Analysis of Variance*. Cambridge University Press, Cambridge, UK.

Weaver, R.W. (ed) (1994) *Methods of Soil Analysis: Part 2: Microbiological and Biochemical Properties*. Number 5 in the Soil Science Society of America Book Series, Soil Science Society of America, Inc., Madison, WI.

Weber, E. (1982) *Air Pollution: Assessment Methodology and Modeling*. Plenum Publishing Corporation, New York.

Weber, U.M. (1997) Dendroecological reconstruction and interpretation of larch budmoth (*Zeiraphera diniana*) outbreaks in two central alpine valleys of Switzerland from 1470–1990. *Trees* **11**, 277–290.

White, G.C. & Garrott, R.A. (1990) *Analysis of Wildlife Radio-Tracking Data*. Academic Press, New York.

Wilson, D.E., Foster, M.S., Cole, F.R., Nichols, J.D. & Rudran, R. (1996) *Measuring and Monitoring Biological Diversity: Standard Methods for Mammals*. Smithsonian Institute Press, Washington, DC.

Winer, B.J., Brown, D.R. & Michels, K.M. (1991) *Statistical Principles in Experimental Design*. McGraw-Hill, Inc., New York.

Zedler, J.B. (1996) Coastal mitigation in southern California: the need for a regional restoration strategy. *Ecological Applications* **6**, 84–93.

CHAPTER 2

Data Management Principles, Implementation and Administration

JAMES W. BRUNT

2.1 Introduction

Effective data management facilitates research and delivers services of use to ecologists. If it constrains science it will be of little use and probably will not persist. Modern data management has its origins in the business sector. However, because of their breadth, complexity and dependence on experimental design, ecological data bear little resemblance to business data. Thus, textbooks in business data management will only be of marginal use for ecological data management. What is attempted here is to compile basic principles and procedures for ecological data management from a variety of sources so that they will be useful to the individual ecologist, as well as the project data manager. With these points in mind, a good place to start is with the philosophy used to guide data management. The role of data management, implementation strategies and data management system components are discussed in latter sections.

2.2 Philosophy

A sound philosophy for research data management in ecology is that it be people-oriented: offering practical solutions to ecologists, placing training and education above technical sophistication and complexity in the computing environment, and providing permanence, ease of access and security for ecological data (Conley & Brunt 1991). A good philosophy also recognizes that scientists want a data management system with a minimum amount of intrusiveness into their limited time and budgets. Adherence to two basic principles can facilitate data management success:
1   start small, keep it simple, and be flexible; and
2   involve scientists in the data management process (see also Chapter 5 (section 5.5.2 and Table 5.3)).

2.2.1 Start small, keep it simple, and be flexible

Avoid unnecessary technological sophistication. Start 'small' and strive for early successes based on existing hardware and software platforms that are familiar to the researcher. Heterogeneity in analytical approaches and computational environments is a fact of life. For instance, most researchers already have access to varying amounts of computing equipment, and most are within reach of a network. Flexibility is critical due to heterogeneity in hardware and software, and differences in ways that ecologists perceive, manage and analyse data. A successful data management system can accommodate the innovative and exploratory nature of science. Good data management is more an approach and a way of thinking than a technological *smörgåsbord* (Conley & Brunt 1991; Brunt 1994; Strebel *et al.* 1994).

2.2.2 Involve scientists in the data management process

A good data management system will take advantage of small-group dynamics and people-oriented solutions. Successful data management systems are dependent on ecologists being integrally involved in the data management process (Conley & Brunt 1991; Brunt 1994; Stafford *et al.* 1994; Strebel *et al.* 1994). This requires face-to-face involvement of scientists in data management decisions; preferably, they are participants in the data management team. Ecologists have the responsibility for defining scientific objectives for the data management system, establishing priorities and defining resource allocation. The data management system should be developed from a research perspective and must reflect the objectives and priorities of the research programme.

2.3 The role of data management

Data and information are the basic products of scientific research. In ecological research, where field experiments and data collections can rarely be replicated under identical conditions, data represent a valuable and, often, irreplaceable resource. For instance, long-term ecological research programmes, both those ongoing and those being developed throughout the world, support research of ecological phenomena occurring over time scales of decades to centuries, periods of time not normally investigated (Magnuson 1990). In long-term ecological studies, retention and documentation of high quality data are the foundation upon which the success of the overall project rests. Twenty years after the fact is too late to discover that data are, for any of myriad reasons, not available or not interpretable for the task at hand. In the short term, scientific results keep the project viable and interesting, thus the data management system must balance the roles of stewardship of data with the need for results.

Ideally, data management can be viewed as a process that begins with the conception and design of the research project, continues through data capture and analysis, and 'culminates' with publication, data archiving and data

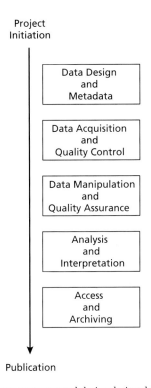

Fig. 2.1 Components of data management and their relationship to research.

sharing with the broader scientific community (Fig. 2.1). Given this definition, all ecologists participate in data management, although they may not be cognizant of doing so! That data management can provide added value to a project's database by assuring archived data are of acceptable quality, and can be retrieved and understood by future investigators is easily seen. Effective data management at the institutional level can also facilitate research and increase productivity through computational support, the development of training programs and support of product-oriented research workshops (Conley 1991; Michener & Haddad 1992).

Data management systems for ecological research have primarily evolved over the last 30 years out of large projects like the International Biosphere Programme (IBP) and the US Long-Term Ecological Research (LTER) Program. The role of data management is continuing to change as research projects are becoming broader and more complex. There are a variety of reasons for implementing a data management system as outlined in Table 2.1. Many future advances in ecology will likely hinge on the ability to integrate diverse computerized data sets, and clearly, carefully considered and applied data management practices are required.

Table 2.1 Reasons for implementing a data management system. (Modified from Risser & Treworgy 1986.)

- Formalize a process to acquire and maintain the products of a research project, including data, so that they may extend beyond the lifetime of the original investigator(s)
- Facilitate the resurrection of currently inaccessible historical data
- Support preparation of data sets for peer-review, publication in data journals, or submission to a data archive
- Provide access to data sets that are commonly used by more than one investigator on a project
- Provide access to data sets by the broader scientific community (e.g. via the WWW)
- Reduce the time and effort spent by researchers in locating, accessing, and analysing data, thereby increasing the time available to synthesize results
- Increase the scope of a research project by facilitating the investigation of broader scale questions that would be otherwise difficult or impossible without the organization of a data management system (e.g. require integration of multiple disparate data sets)
- Incorporate data from automated acquisition systems into ongoing analytical efforts such as ecological modelling.

2.4 Data management implementation

As previously noted, all ecologists participate in data management as a normal part of their research. In many cases, data management by an individual scientist consists of those activities required to prepare data for analysis (e.g. data entry, quality assurance, processing). Attention to data management often ends abruptly with publication of the results. More formalized and long-term data management often does not take place until there is a perceived need (e.g. a project becomes too complex) or it is mandated by a funding agency (also see Table 2.1). When this happens it is necessary for the investigator, project or institution to establish data management guidelines and implement a data management system.

There is no one right way to manage ecological data, but a logically organized structure for the data management system is clearly important. The design of an effective data management system depends on considerable forethought and planning to meet and balance several fundamental requirements or objectives (Stafford *et al.* 1986; Michener & Haddad 1992; Briggs & Su 1994; Brunt 1994; Lane 1997). The primary goal of a system is to provide the best quality data possible within a reasonable budget. Achieving high quality data may entail consulting with technicians and investigators on design and collection of data, assuring that entry errors are found and corrected, providing consistent and acceptable documentation, and supporting the feedback of summary information to investigators for scrutiny. A second system requirement is facilitating access to data by investigators. Third, parallel in importance to access, is providing short-term and long-term security for data through data archiving. Archival storage involves various activities (may include data publication) that are designed to protect the data through natural or human-made disasters. Finally, the fourth system requirement is providing computational support for the investigator's analysis of collected data, including, but not limited to, quality assurance. Additional efforts may include providing routine data analyses for investigators, supporting electronic communication and facilitating the timely dissemination of research results.

The protocols and computational infrastructure required to achieve these goals vary considerably within the research community and can range from relatively simple to extremely sophisticated implementations. The research programme must balance resources expended on data management with those expended on data collection and analysis activities to best meet research objectives. Technology affects the way data management is performed but should not affect the principles that are applied. Furthermore, the management of research data should be achieved through practical solutions that will withstand change.

2.5 Data management system components

Research data management in ecology, as in other disciplines, can involve acquiring, maintaining, manipulating, analysing and archiving data, as well as facilitating access to both data and results (Michener 1986; Conley & Brunt 1991). From an organizational perspective, the activities also require attention to infrastructure, personnel, policies and procedures, which *in toto* constitute the working data management system.

Six components or activities are integral to implementation of a data management system.

1 An inventory of existing data and resources will have to be compiled, and priorities for implementation set (Gurtz 1986; Michener 1986).

2 Data will have to be designed and organized by establishing a logical structure within and among data sets that will facilitate their storage, retrieval and manipulation.

3 Procedures will be required for data acquisition and quality assurance and quality control (QA/QC) (also see Chapter 4).

4 Data set documentation protocols, including the adoption or creation of metadata content standards and procedures for recording metadata, will need to be developed (Michener *et al.* 1997; also see Chapter 5).

5 Procedures for data archival storage, and maintenance of printed and electronic data will have to be developed (see Chapter 6).

6 Finally, an administrative structure and procedures will have to be developed so responsibilities are clearly delineated.

2.5.1 Inventory of existing data sets

It is essential to have an inventory of data and resources that includes not only past and present, but also planned data sets and resources. Research activities and programmes; types, amounts, and forms of the data; and staff, money and facilities should all be included. After the inventory is completed, decisions will have to be made regarding the objectives for each data set, and a list with as much supporting information as possible should be made available to the project investigators. This list will be invaluable for the beginning data management system and will be requested, used and revised countless times.

2.5.2 Data design, organization and manipulation

Data design

Ecologists can avoid many potential difficulties in field sampling and subsequent data analyses if sufficient thought is given to designing data sets *prior* to collecting data. In most cases, the preliminary data design should reflect the experimental design in use. Some decisions about data design are necessary before data are collected in order to produce field and laboratory data sheets. The completed design can be transferred directly to data entry tools to aid in data collection, to facilitate analysis by statistical software and to support metadata development and structure the data set for archiving (Fig. 2.2; also see Chapters 5 & 6).

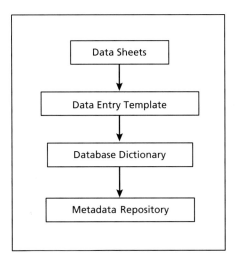

Fig. 2.2 An example of the flow of attribute information through a typical project data management system. Informed design decisions made prior to collecting data can alleviate data set organization problems later.

DATA MANAGEMENT PRINCIPLES

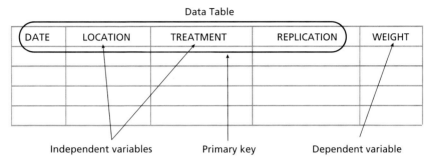

Fig. 2.3 Example of the relationship between experimental design and data table.

Data design refers to the 'logical' organization of data into tables. Most ecological data are collected under rigorous experimental designs and are therefore suitable for incorporation into statistical tables (e.g. Fig. 2.3). Statistical tables are relational tables designed to reflect the dependent and independent variables of the experiment. If an experiment involves multiple environmental factors (e.g. observations of biomass and precipitation), multiple tables may be needed for efficient structuring and retrieval of data (including maintenance of proper relationships and integrity rules). In such cases, one or more relational attributes or key attributes should be common across tables (Fig. 2.4). The tables can later be joined by the relational attributes when they are converted into digital data files, databases or spreadsheets for analysis. The process of designing the precise structure of data tables for implementation in a database management system (DBMS) is called normalization. Application of this design process to ecological data requires *detailed knowledge of the data* to avoid costly mistakes (see Chapter 3).

Organization of data sets

Putting existing data into a computer 'where they belong' is often the first perceived need of a project. Commercial DBMS software may be inappropriate for initial file organization and management. One way for beginning projects to approach data set organization is with a file management system. In keeping with the 'small and simple' approach, a beginning project might organize data sets around the file handling capabilities of operating systems such as Windows NT and UNIX. Using operating system tools and shell programming for application and database programs eliminates the overhead of maintaining

32   CHAPTER 2

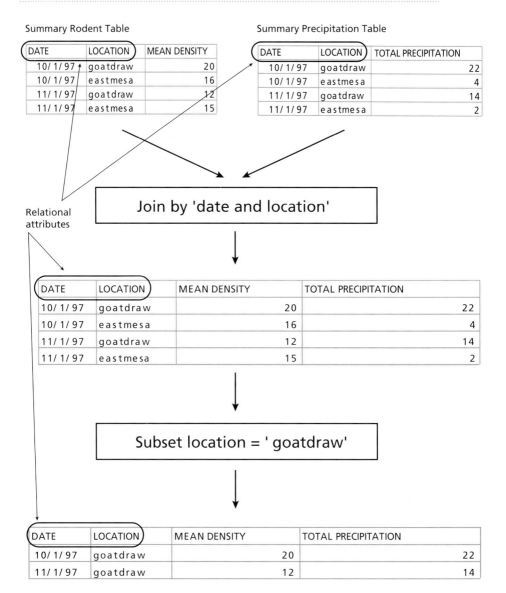

Fig. 2.4 Multiple data tables for a single project, reflecting the use of relational attributes to join and subset data tables.

a commercial DBMS (Manis *et al.* 1988) until it is needed. This approach is referred to as a Data File Management System (DFMS) (Hogan 1990). Data and metadata can be stored in computer files in specific areas of the hierarchical directory structure of the computer system. Analytical and management tools can also be stored in discipline-specific areas of the same file system. Certain areas of the directory structure can be used for data entry, manipulation

DATA MANAGEMENT PRINCIPLES 33

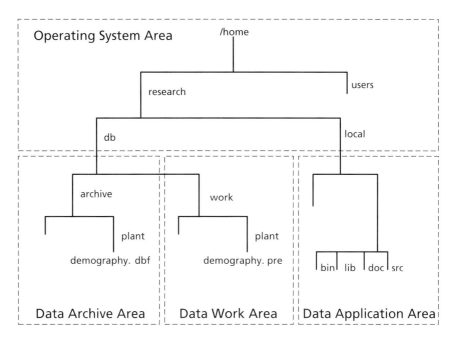

Fig. 2.5 Example of the use of computing system directory structure in the data management system. The dashed lines indicate physical separation on hard disks while the solid lines represent virtual links between directories.

and construction of the archive files (Fig. 2.5). A well-structured DFMS can meet many of the initial DBMS objectives (Hogan 1990). However, a DBMS can force the data manager to be more thoughtful about data integration, and file-based systems may be more prone to data entropy.

DBMS packages were developed for business applications where all data are logically related and constitute a single logical database. In large, diverse research projects there can be many logical databases. Each ecological database may be treated as a unit of data collected under the same monitoring or experimental design by the same sampling methods. In practice, commercial DBMS software is often necessary and can be made to work in numerous scientific applications, but its implementation needs to be done carefully and with much forethought about the need for and use of the system (see Chapter 3).

Data manipulation and maintenance

Managing data involves the use of special software tools to store, manipulate and maintain scientific data. There are myriad software tools available to the data manager, some specialized and some very general in scope. Some software packages such as SAS and S+ are designed specifically for the manipulation and analysis of scientific data, while others such as Foxpro, Oracle and

MS SQL Server are designed to support general database management system functions. Accounting tools like spreadsheet software (Excel, Lotus 1-2-3) are readily available and can perform limited manipulation and analysis functions. However, caution should be exercised when relying on these tools because, by default, they may not maintain internal record consistency, i.e. each column can be manipulated independently of the row. Internal record consistency is a standard feature of more specialized analytical tools like SAS and database management system (DBMS) software. One of the most important decisions by the data manager will be whether to invest time and money in a DBMS.

The DBMS can be the heart of an integrated data management system, but the decision will depend on many factors, including the types and complexity of data relationships, level and type of data access required and the intended uses of the data (see Chapter 3). Geographic information systems (GIS) represent a special type of DBMS that combines spatial mapping and analytical capabilities with relational database functions. These systems have the capability to form the hub of a data management system for projects that are predominantly spatially oriented. GIS has a steep learning curve for taking full advantage of its features as a data management tool. It should be viewed as part of the overall data management system, particularly if the research has a strong spatial component. Considerations, including the use of accepted practices in storing and documenting coordinates, should be made in all data management systems for the potential use of data in GIS. Johnston (1998) discusses concepts and procedures for implementing GIS for ecological research in detail.

The production and delivery of derived data products, and advanced query, integration and analysis systems, may require the implementation of an Internet-based information system. An information system expands the capabilities of data and database management systems by providing additional integrative services and access. A conceptual example is the development of databases that integrate other databases and then provide the integrated product via World Wide Web (WWW) servers (e.g. Henshaw *et al.* 1998). The popularity of the WWW has resulted in numerous sources of information designed to aid the developer in putting together an Internet information service (Liu *et al.* 1994).

2.5.3 Data acquisition and QA/QC

Ecological data are often collected first on paper and subsequently transferred to a computer for analysis and storage. Paper may provide some advantages in terms of longevity and ease of use, but it does not work well under some environmental conditions, and provides few alternatives for processing until the data are transferred to computerized form. Increasing numbers of researchers are collecting data directly in the field on small computers, where environmental factors allow. There is also increasing use of automated data collection instruments that record data directly to a computer. Typically, the fewer number of times data are transferred from one form to another, the fewer errors will be introduced. Transfer of data from one form to another should ideally occur only once, completing appropriate QA/QC during that process (see Chapter 4).

Paper data forms

Paper is likely to continue to be a primary data acquisition tool, therefore it is important to exercise good judgement in the use of paper data forms. Always use acid free paper to prevent fading and subsequent loss of data. Depending on conditions it may be necessary to use paper and writing devices that can withstand moisture, dust, and other extreme environmental conditions. Data sheets should be bound in some way both in use and in storage because of the likelihood that one or more sheets might become separated and lost from the rest. Each page should contain basic information about the study it will be used for prior to collection. Each page should be numbered and contain a place for relevant metadata and comments like date, collectors, weather conditions, etc. As much as possible, each page should reflect the design structure of the data set. This single requirement will greatly reduce entry errors and make QA/QC procedures much easier because of the 1 : 1 correspondence of attributes.

Tape recorders

Many researchers are using small hand-held microcassette tape recorders as an alternative to paper in the field. Later, recorded observations are transcribed to paper or entered into computer files in the laboratory under more favourable computational and environmental conditions. As with any technological solution, there are drawbacks. These include battery and tape maintenance, low environmental tolerance, and risk of failure. However, tape recorders can provide a high quality, efficient method of collecting data that can be easily operated by a single person in the field.

Hand-held computers

Small notebook, or palmtop, computers are being used more frequently to collect data in the field. Using computers at the point of collection has the advantage of incorporating QA/QC checks while there is still an opportunity for resolution and correction. This method of data acquisition probably provides the highest quality data when combined with point-of-entry data quality checks. However, it also carries a hefty price tag in comparison to other methods, and is subject to narrow environmental constraints (e.g. heat, moisture and dust). Data collected on hand-held computers should be backed up routinely and frequently because of the high risk of equipment failure.

Automated data acquisition systems

Many data today are being collected using instruments that have their own data acquisition systems. Probes and radiometers, as examples, automatically log data; these data are subsequently downloaded or transferred to another computer for processing. Similarly, the use of small data loggers is widespread. These simple computers collect data continuously from a variety of environmental sensors. Data loggers are an efficient method of collecting continuous sensor data, but they must be routinely downloaded as physical memory is usually limited, and they can require considerable programming and electronic expertise.

Coupled with modems and in combination with radio or telephone communication systems, data loggers can provide hands-off transfer of data to a central acquisition point. This method can be very efficient where the data are needed in a routine frequency and where there are difficulties in acquiring data because of the remoteness, number and/or inaccessibility of the sites. Again, as with all technological solutions, there are environmental constraints as well as power and maintenance requirements that must be considered.

Most modern lab instruments are equipped with computer interfaces allowing the use of data acquisition software on a personal computer to receive and process the data. This greatly reduces errors of transcription and saves time and energy.

QA/QC

QA/QC mechanisms are designed to prevent the introduction of errors into a data set, a process known as data contamination. There are two fundamental types of errors that can occur in data, those of commission and those of omission (Kanciruk *et al.* 1986). Errors of commission include incorrect or inaccurate data in a data set resulting from a variety of sources, such as malfunctioning instrumentation and data entry and transcription errors. Such errors are common and are relatively easy to identify. Errors of omission, on the other hand, may be much more difficult to identify. Errors of omission frequently include inadequate documentation of legitimate data values, which would affect the way a given data value is interpreted.

Quality control (QC) procedures can be very effective at reducing errors of commission. Typically, quality control refers to mechanisms that are applied in advance, with a priori knowledge, to 'control' data quality during the data acquisition process. These procedures are performed during data collection and data transcription to prevent corruption or contamination. Data sheets that are partially filled out for plot and place names to prevent the introduction of illegal data are a simple example of a quality control mechanism. Point-of-entry software that refuses to allow an 'N' to be entered where only an 'M' or 'F' is valid is another example. More advanced point-of-entry software checks spelling and compares entries to look-up tables.

Quality assurance (QA) procedures are used to identify errors of omission and commission. Quality assurance mechanisms can be applied after the data have been collected, transcribed, entered in a computer and analysed. A much more broad-reaching application, quality assurance can be as simple as plotting data values to look for outliers. More advanced mechanisms of quality assurance involve statistical methods for describing and analysing a data set or its results and assigning probabilities to the quality of data. Data quality assurance is discussed in more detail in Chapter 4.

Combined QA/QC for ecological data includes four activities that range from relatively simple and inexpensive to sophisticated and costly.
1   Defining and enforcing standards for formats, codes, measurement units and metadata.
2   Checking for unusual or unreasonable patterns in data.
3   Checking for comparability of values between data sets.
4   Assessing overall data quality.

Most QA/QC is typically in category 1. This most basic element of QA/QC begins with data design and continues through data acquisition, metadata development and preparation of data and metadata for submission to a data archive. Examples of QA/QC for each of these stages are listed in Table 2.2.

A typical situation would involve QA/QC procedures in the process of transferring data from the field into a database management system. Comprehensive QA/QC includes data entry with point-of-entry checks, data verification and correction, and data validation through review by qualified scientists (Fig. 2.6). This process from raw data to verified data to validated data is referred to as data maturity (Strebel *et al.* 1994) and implies an increasing confidence in the data quality through time.

2.5.4 Data documentation (metadata)

Metadata represent the structural, logical, and methodological documentation of a data set—literally, information about data (Michener *et al.* 1997). Consider the examples in Fig. 2.4. Without supporting metadata, these data would be meaningless. Metadata for these data sets could easily be more extensive and complex than the data themselves. Where data end and metadata begin is often the subject of much discussion. One conceptual model of the relationship

Table 2.2 Quality assurance and quality control procedures that are associated with data design, data acquisition, metadata development and data archival phases in a comprehensive data management system.

Quality assurance and quality control (QA/QC)	Design	Acquisition	Metadata	Archive
Check that data sheets represent experimental design	✓			
Check that measurement units are defined on the data sheet	✓			
Check that attribute names meet project standards	✓			
Check that date, site, and coded values meet project standards	✓			
Check that attribute names and descriptions are provided	✓			
Check that data are complete		✓		
Check that data entry procedures were followed		✓		
Check that data include time, location, and collector(s)		✓	✓	✓
Check that measurement data are within the specified range		✓		
Check that data values or codes are represented correctly		✓		
Check that data are formatted correctly for further use		✓	✓	✓
Check that data table attribute names are reasonable		✓	✓	✓
Check that data table design reflects experimental design		✓	✓	✓
Check that values for each attribute are represented one way		✓	✓	✓
Check that errors and corrections are recorded		✓	✓	✓
Check that metadata are present			✓	✓
Check metadata for content (accuracy and completeness)			✓	✓
Check that data dictionary is present and accurate			✓	✓
Check that measurement units are consistent		✓	✓	✓
Check that data and metadata are complete				✓

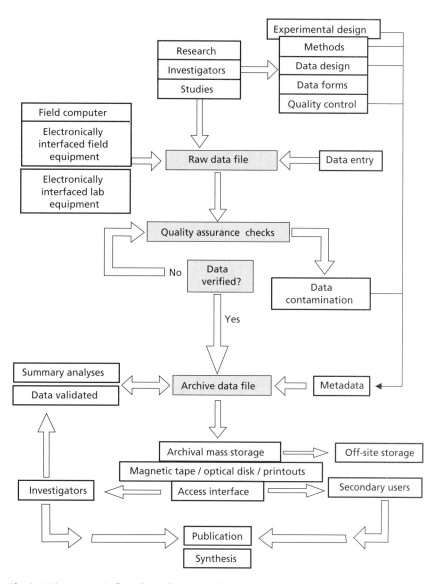

Fig. 2.6 Diagrammatic flow chart of a generic data processing and archival storage procedure.

of metadata to data equates the metadata for a data set to an object in a key field of the record that contains the measurement or observation (Brunt 1994). All the metadata necessary to understand an observation would be compressed into a data object that would fit under a single attribute in a table. This simple model for understanding the importance of metadata assumes the experimental design, sampling methods and other supporting documentation are a fundamental and logical component of each data record (Fig. 2.7). Thus, an

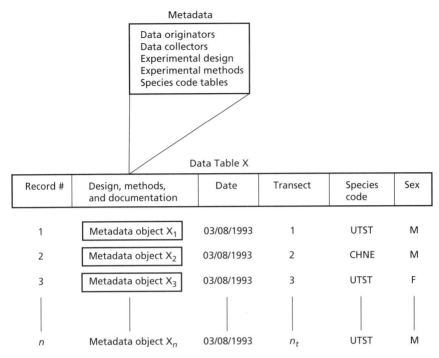

Fig. 2.7 Conceptual relationship of the design, methods, and metadata (represented as a single object) to the observation of ecological phenomenon. The objects become a logical and obligate part of the primary key to each observation.

observation and its supporting documentation constitute a single logical record and make that observation unique. In current practice, this model would be impractical; however, it highlights the principle that metadata should ideally be stored *in an integrated manner* with the data set for preservation (Conley & Brunt 1991; Brunt 1994; Ingersoll *et al.* 1997; Michener *et al.* 1997). It is likely that further developments in metadata containers (Daniel & Lagoze 1997) and object-oriented database management systems (OODBMS) will make this model feasible. Issues related to metadata that should be considered in any data management system are discussed more fully in Chapter 5.

2.5.5 Archival storage and access

Data stored in a computerized information system may be of tremendous value not only to the project but to the broader ecological community as well. The value is determined not only by what is stored, but *how* it is stored. The type of storage device, the format of the stored data and types of access available to those data have a significant influence on the ultimate value of the stored data. Ideally, once entered and verified, data are transferred to an archive file or format. The archive file then becomes the reference version of the data, regardless of whether it exists locally or in a formal data archive. All subsequent work should be done to the archive file or to a copy of the archive file that will replace the original. This process helps to control the problem of proliferating 'offspring' files containing different versions of the data.

Once data are in an archive they can easily be made accessible to the investigators via an electronic network. Simple access can be provided through a web browser interface that allows the user to download the data in a variety of formats compatible with available analytical tools. In most cases, data should be archived online, with the possible exception of image data, which may, even today, consume inordinate amounts of disk storage space and be impractical to download. The online data should be copied and saved to tape or optical disk and placed in two or more locations off the premises to protect against disasters. It is important to keep these copies, as well as the on-site copy, up-to-date! Data archiving is discussed in more detail in Chapter 6.

2.5.6 Data administration

Data can originate from a variety of sources and be stored in many locations in different formats. Heterogeneity of this kind makes administration of the data management system an important task. Although most individual researchers manage their own data, large projects may have one or more individuals dedicated to the task of data management. The function of data administration is to coordinate the storage and retrieval of computerized information and oversee the implementation of data management procedures and policies in larger projects. Establishing policies that protect the project data is an important activity of data administration. Policies should include procedures for backup and recovery, as well as access. Effective data management in larger projects can provide added value to the data by meeting project goals for data quality, data access, data security and project support, as well as meeting the individual researcher's need for autonomy.

The scope of data administration activities is dependent on the degree of centralization of the data management system and the degree to which investigators manage their own data. A project with a large number of investigators distributed around the country at various institutions will emphasize a different set of data management functions than those for projects where all the investigators are located at the same institution. The more distributed the project, the more important integration becomes. Decentralized projects may have no data management staff and little coordination or integration of data management activities. In this case, the responsibility for these activities rests with the individual investigators.

An important and complex data administration task is to establish policies and procedures that define the level of data management involvement in various aspects of research. For the data administrator, research and development

> **Box 2.1 Generic data access policy**
>
> Data collected under the auspices of this project are available upon request to qualified scientific interests that agree to cite the data and source appropriately. Ancillary data sets from research performed by investigators on the project are also available once the investigator has released the information, or 2 years after the termination of the study, to give the investigator adequate time to present findings in the scientific literature. For requests from outside the scientific community it may be necessary to charge handling and processing fees according to our request policy.

of software tools are often necessary, and may significantly enhance the efficiency of data management effort devoted to project support. Handling requests and training personnel can also present particular challenges to the system. Future maintenance of the project data will require strategies for attracting and retaining data management staff.

Data access policy

Guidelines for data access establish the means for determining who should be provided access to specific data, as well as the conditions and mechanisms for access. There are many ways a data management system can facilitate data sharing but it is first necessary to establish an institutional or project-level policy. However, access policies can be controversial, and may have to be dealt with on a case-by-case basis. For instance, some scientists may not be inclined to share data despite the fact that benefits to the data contributor often outweigh the costs (Porter & Callahan 1994). A generic data access policy, as an example, is provided in Box 2.1.

Supporting research

Defining the level of involvement of data management in supporting research requires clearly delineated functions and establishment of priorities. The example in Table 2.3 divides the functions into primary, secondary and tertiary support levels (after Brunt 1994). Primary support includes database administration, security administration, quality assurance and technical consultation, while secondary support includes computer system administration and data analysis, and tertiary support is the development of tools for various aspects of the research programme. The priorities of the individual research programme would establish what functions receive top priority levels. For example, many projects might include computer system administration under primary support while others might depend on outside sources for this service.

Table 2.3 Research support functions of data management.

Support priority levels	Functions	Comments
Primary support	Database administration	Includes creation and maintenance of data sets and database objects
	Security administration	Includes backup and recovery, archiving and storage space management
	QA/QC	Quality assurance/quality control mechanisms
	Technical consultation	Includes data design and training
Secondary support	Computer system administration	Hardware, operating system, commercial and in-house software
	Data manipulation and analysis	For additional quality assurance and/or publication
	Graphics and text production	For publications and presentations
	Multimedia	For presentations, including WWW
	Data support for individual investigator projects	Data QA/QC, manipulation and analysis for publication and presentation
Tertiary support	Tool and technique development	For data QA/QC, manipulation, analysis, access and archiving
	Tool support for individual investigator projects	For data QA/QC, manipulation, analysis, access and archiving

It is anticipated that 75% of available time would be spent in the primary support category (Pasley 1991). As priority decisions are made, it is important to consider the long-term implications of not providing some services for an ever-increasing volume of data.

Project support has to be scheduled based on the availability of resources and time for completion. Administration must track actual support loads to determine real time available for investigator projects (Pasley 1991). A process of feeding resources back into tool development increases the efficiency of activities requiring a lot of support (e.g. data entry and verification) and subsequently increases the amount of resources that can be dedicated to the support of investigator projects. Reduction of the time requirement for getting the data ready for analysis should be one of the main goals of data administration.

An issue affecting each category is the 'competition' for support. This competition necessitates that each of the data management personnel should have specific goals for the amount of time allocated to the support areas. First priority must be given to the primary support functions, like maintenance of project data. Subsequently, data management personnel, in consultation with the data administrator and the project investigators, must make decisions about the level and types of requests that can be handled given the goals of the project and current activities.

Handling requests

A substantial amount of data management time is spent handling requests from both inside and outside the project. It doesn't take many requests for a small data management system to be overwhelmed. As the project grows, it is important to implement procedures to deal with these requests (Michener & Haddad 1992). This accomplishes two things:

1 it relieves the data management personnel of some of the burden of juggling an enormous number of requests; and
2 it allows them to work more productively toward completion of the requests that are assigned.

Requests to data management staff include requests for data, software development, analysis and technical assistance. The system of handling requests should not hinder anyone's access to data or analysis tools; it should provide a fair and systematic way to deal with requests and track their progress. Upon receiving a request, an individual can handle the request personally, juggling his or her own time and resources, or the request can be passed into a 'request mill,' where it is evaluated, assigned, and given a priority (Fig. 2.8). Including a mechanism for tracking requests as well as progress is important. Prior to the use of tracking procedures, data management staff will almost always underestimate the amount of time spent providing technical assistance and handling data requests. Efforts should be made to maintain adequate levels of assistance, but a project should also implement training opportunities so individuals can develop greater self-sufficiency.

In addition to these internal requests, data management personnel may receive requests from agencies, organizations and individuals outside the project. Policies must be established for handling these requests which, by necessity, may be treated differently. Policies may include request forms, waivers and a pay-as-you-go pricing schedule. This last item may be critical because most projects do not have funds to support filling requests by individuals or agencies not directly associated with the projects. If the project receives a large number of requests for data, a portion of the costs associated with providing access can be recovered through retrieval fees or data licenses.

Data management personnel

Overall, management of project data sets is the function that will consume the most time and resources in data management. In large projects, this effort is headed by an individual with the title data manager, information manager or data administrator. One common mistake is the hiring of a database manager

DATA MANAGEMENT PRINCIPLES   45

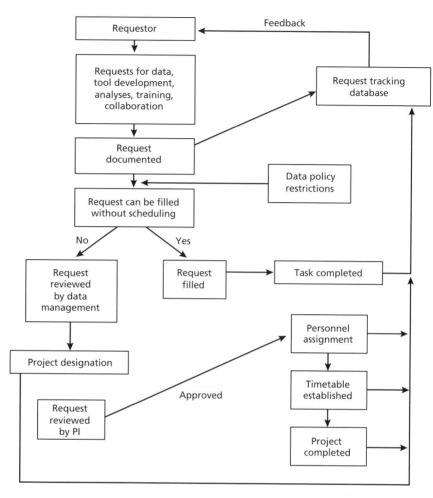

Fig. 2.8 Diagrammatic chart of a data management request handling procedure.

that everyone 'expects' to be a data manager—there is an important distinction between the two. A database manager deals with the technical issues involved in storage and retrieval inside the complex environment of database management system (DBMS) software. A data manager, data administrator or information manager deals with administration and management of all the project data, not just those data stored in a DBMS.

Data administrators currently face several major challenges in keeping up with project growth, particularly in regard to data management personnel. Staff gain skills that are in high demand in the private sector; student personnel come and go as they complete their degrees. Personnel turnover is inherent in the academic environment, and consequently represents a significant challenge to developing and maintaining a data management system.

> **Box 2.2 Data management keys to success**
>
> • Start small and keep it simple—building on simple successes is much easier than failing on large inclusive attempts.
> • Involve scientists—ecological data management is a scientific endeavour that touches every aspect of the research programme. Scientists should be involved in the planning and operation of a data management system.
> • Support science—data management must be driven by the research and not the other way around, a data management system must produce the products and services that are needed by the community.

There are two basic approaches for dealing with personnel turnover. First, and ideally, salaries and benefits should be competitive with the private sector; good personnel require incentives to stay. Second, redundancy in skill and job tasks should be built into hiring strategies. Redundancy gives the system the ability to continue to operate when personnel come and go.

2.6 Conclusion

There are myriad tools and techniques available to support data management. Each project's data management system functions within a unique set of constraints, so strict application of many of the methods described in this chapter and in other chapters in this volume may be impractical. However, some basic guidelines will help most data managers and project managers along the road to success for their project (see Box 2.2).

2.7 References

Briggs, J. & Su, H. (1994) Development and refinement of the Konza Prairie LTER research information management program. In: *Environmental Information Management and Analysis: Ecosystem to Global Scales.* (eds W.K. Michener, J.W. Brunt & S.G. Stafford), pp. 87–100. Taylor and Francis, Ltd., London.

Brunt, J.W. (1994) Research data management in ecology: a practical approach for long-term projects. In: *Seventh International Working Conference on Scientific and Statistical Databases.* (eds J.C. French & H. Hinterberger), pp. 272–275. IEEE Computer Society Press, Washington, DC.

Conley, W. (1991) An institute for theoretical ecology? Part IV: 'Computational workshops': a planned activity for theoretical ecology. *Coenoses* **6**, (2), 113–120.

Conley, W. & Brunt, J.W. (1991) An institute for theoretical ecology? Part V: Practical data management for cross-site analysis and synthesis of ecological information. *Coenoses* **6**, (3), 173–180.

Daniel, R., Jr. & Lagoze, C. (1997) Extending the Warwick Framework: From Metadata Containers to Active Digital Objects. *D-Lib Magazine.* ISSN 1082–9873. Volume: November, 1997.

Gurtz, M.E. (1986) Development of a research data management system. In: *Research Data*

Management in the Ecological Sciences. (ed. W.K. Michener), pp. 23–38. University of South Carolina Press, Columbia, SC.

Henshaw, D.L., Stubbs, M., Benson, B.J., Baker, K., Blodgett, D. & Porter, J.H. (1998) Climate database project: a strategy for improving information access across research sites. In: *Data And Information Management in the Ecological Sciences: A Resource Guide*. (eds W.K. Michener, J.H. Porter & S.G. Stafford), pp. 29–31. LTER Network Office, University of New Mexico, Albuquerque, NM.

Hogan, R. (1990) *A Practical Guide to Database Design*. Prentice-Hall, Englewood Cliffs, NJ.

Ingersoll, R.C., Seastedt, T.R. & Hartman, M. (1997) A model information management system for ecological research. *BioScience* **47**, 310–316.

Johnston, C.A. (1998) *Geographic Information Systems in Ecology*. Blackwell Science, Oxford, UK.

Kanciruk, P., Olson, R.J. & McCord, R.A. (1986) Quality control in research databases: The US Environmental Protection Agency National Surface Water Survey experience. In: *Research Data Management in the Ecological Sciences*. (ed. W.K. Michener), pp. 193–207. University of South Carolina Press, Columbia, SC.

Lane, A.M.J. (1997) The UK Environmental Change Network Database: An integrated information resource for long-term monitoring and research. *Journal of Environmental Management* **51**, (1), 87–105.

Liu, C., Peek, J., Jones, R., Buus, B. & Nye, A. (1994) *Managing Internet Information Services*. O'Reilly and Associates Inc., Sebastapol, CA.

Magnuson, J.J. (1990) Long-term ecological research and the *invisible* present. *BioScience* **40**, 495–501.

Manis, R., Schaffer, E. & Jorgensen, R. (1988) *UNIX Relational Database Management Application Development in the UNIX Environment*. Prentice-Hall, Englewood Cliffs, NJ.

Michener, W.K. (1986) Data management and long-term ecological research. In: *Research Data Management in the Ecological Sciences*. (ed. W.K. Michener), pp. 1–8. University of South Carolina Press, Columbia, SC.

Michener, W.K. & Haddad, K. (1992) Database administration. In: *Data Management at Biological Field Stations and Coastal Marine Labs: Report of an Invitational Workshop*. (ed. J. Gorentz), pp. 4–14. W.K. Kellogg Biological Station, Michigan State University, Hickory Corners, MI.

Michener, W.K., Brunt, J.W., Helly, J., Kirchner, T.B. & Stafford, S.G. (1997) Nongeospatial metadata for the ecological sciences. *Ecological Applications* **7**, (1), 330–342.

Pasley, R. (1991) Managing DBAs in the turbulent 1990s. *Database Programming and Design* **4**, (2), 34–41.

Porter, J.H. & Callahan, J.T. (1994) Circumventing a dilemma: historical approaches to data sharing in ecological research. In: *Environmental Information Management and Analysis: Ecosystem to Global Scales*. (eds W.K. Michener, J.W. Brunt & S.G. Stafford), pp. 194–202. Taylor and Francis, Ltd., London.

Risser, P.G. & Treworgy, C.G. (1986) Overview of ecological research data management. In: *Research Data Management in the Ecological Sciences*. (ed. W.K. Michener), pp. 9–22. University of South Carolina Press, Columbia, SC.

Stafford, S.G., Alabach, P.B., Waddell, K.L. & Slagle, R.L. (1986) Data management procedures in ecological research. In: *Research Data Management in the Ecological Sciences*. (ed. W.K. Michener), pp. 93–114. University of South Carolina Press, Columbia, SC.

Stafford, S.G., Brunt, J.W. & Michener, W.K. (1994) Integration of scientific information management and environmental research. In: *Environmental Information Management and Analysis: Ecosystem to Global Scales*. (eds W.K. Michener, J.W. Brunt & S.G. Stafford), pp. 3–19. Taylor and Francis, Ltd., London.

Strebel, D.E., Meeson, B.W. & Nelson, A.K. (1994) Scientific information systems: a conceptual framework. In: *Environmental Information Management and Analysis: Ecosystem to Global Scales*. (eds W.K. Michener, J.W. Brunt & S.G. Stafford), pp. 59–85. Taylor and Francis, Ltd., London.

CHAPTER 3

Scientific Databases

JOHN H. PORTER

3.1 Introduction

A scientific database is a computerized collection of related data organized so as to be accessible for scientific inquiry and long-term stewardship. Scientific databases make possible the integration of disparate data as well as uses of data in new ways, often across disciplines. The National Research Council publication *Bits of Power* notes a recent trend: 'The need for scientists to adapt to conducting research with data that come in rapidly increasing quantities, varieties, and modes of dissemination, frequently for purposes far more interdisciplinary than in the past' (NRC 1997). To meet the challenges created by this trend will require unprecedented data exchange among scientists to support integrative analyses; such integration may best be facilitated by the development of scientific databases.

There are several advantages to developing and using scientific databases. The first is that databases lead to an overall improvement in data quality. Multiple users provide multiple opportunities for detecting and correcting problems in data. A second advantage is cost. Data generally cost less to save than to collect again. Often, ecological data cannot be collected again at any cost because of the complex nature of poorly controlled factors, such as weather, that influence population and ecosystem processes.

However, the primary reason for developing scientific databases must be the new types of scientific inquiry that they make possible. Gilbert (1991) discusses the ways databases and related information infrastructure are leading to a paradigm shift in biology. Nowhere has this been more evident than in the genomic community, where the creation of databases and associated tools have facilitated a tremendous increase in our understanding of the relationship between the genetic sequences and the actions of specific genes.

Ecology awaits a similar renaissance, brought on through improvements in databases and data communication. Specific inquiries requiring databases include long-term studies, which depend on databases to retain project history; syntheses, which often combine data for a purpose other than that for which the data were collected; and integrated multidisciplinary projects, which depend on databases to facilitate sharing of data. Databases make it possible to integrate diverse data resources in ways that support decision-making processes.

3.2 Challenges for scientific databases

Developers of scientific databases face challenges that are different from those experienced by most business-oriented databases (Pfaltz 1990; Robbins 1995). As noted by Robbins (1995), the technologies supporting business databases emphasize data integrity and internal consistency. It would not do to have two disparate estimates of hours worked when paychecks are being issued. However, scientific databases may contain observations of the same phenomenon that are inconsistent, resulting from differences in methodology and measurement imprecision or even different models of the physical processes underlying the object of study. Additionally, with the exception of correction of errors, scientific data are seldom altered once placed in a database. This contrasts with business data where an account balance may be altered repeatedly as funds are expended. For this reason, several authors (Cincowsky *et al*. 1991; Robbins 1994, 1995; Strebel *et al*. 1994, 1998; Meeson & Strebel 1998) have proposed a model for scientific database development analogous to publication of scientific information rather than the traditional, business database model.

Science means asking *new* questions. Scientific databases thus need to be adaptable so that they can support new kinds of queries. In most business-oriented databases, the focus is on development of standardized queries. This month's sales report is similar in form (although not content) to last month's report. Business-oriented database software has many features that aid in the production of standardized reports. However, for ecological data, queries and analyses are done in a more iterative way that shifts the focus of the analysis with each iteration to look for new relationships within a given data set or new linkages with other data sets.

The biggest challenge to developing useful scientific databases is dealing with data diversity. The volume and complexity of scientific data vary widely (Fig. 3.1). Some types of data have a high volume but are relatively homogeneous. An example of this is image data from satellites. Although each image may require hundreds of megabytes to store, the storage formats of the data are relatively standardized and hence require relatively little metadata for use. In contrast, certain types of manually collected ecological data have an extremely small volume but require extensive documentation to be useful. For example, deep soil cores are very expensive to obtain, so data are usually restricted to a few cores. However, these cores undergo many different analyses: examining the density, mineral content, physical characteristics, biological indices, isotopic ratios, etc. Each of these analyses needs to be well documented, thus the requisite metadata can exceed the volume of numerical data by several orders of magnitude. Some data are both high in volume and complex. Geographical Information System (GIS) data layers can be very high in volume (depending on the resolution of the system) and require metadata

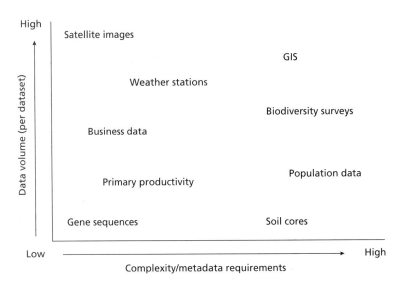

Fig. 3.1 Relationship between data volume and complexity.

that cover not only the actual data collection but the processing steps used to produce it (Federal Geographic Data Committee (FGDC) 1994).

In some areas of science (e.g. genomics, water quality) there is wide agreement on particular types of measurements and the techniques for making them. However, in other areas (e.g. the measurement of primary productivity), there is relatively little agreement and the types of data collected are much more diverse. Standards are most common in 'mature' areas of inquiry. In less mature areas of science, experimentation with methodologies is a necessary part of the scientific process. Sometimes, there is a convergence in methodologies, leading to the informal adoption of emergent standards, which may be adopted subsequently as formal standards (also see Table 1.1 and related discussion in Chapter 1). This process is especially rapid where there are a limited number of specialized instruments for making measurements. Conversely, the standardization process is especially difficult where a methodology needs to operate across a range of ecological systems. Techniques developed for aquatic systems may be impossible to apply in forested systems.

The challenge of diversity extends to users as well. Scientific users have different backgrounds and goals that need to be supported by the database. Moreover, the user community for a given database will be dynamic as the types of scientific questions being asked change and new generations of scientists use the database (Pfaltz 1990).

Today, scientific databases require a long-term perspective, which is foreign to many other types of databases. Indeed, centuries-old data are particularly valuable as they allow us to assess changes that would otherwise be invisible to us (Magnuson 1990). A frequently cited goal for an ecological database is that data be both accessible and interpretable 20 years in the future (Justice *et al.* 1995 extend this goal to 100 years).

3.3 Developing a scientific database

The development of a database is an evolutionary process. A database will serve a dynamic community of users during its lifetime and it needs to change to meet changing needs. For this reason, it is important to avoid the temptation of creating the 'ultimate database'. In creating a database, four questions need to be asked. The first is 'Why is this database *needed*?' Not all data are important enough to warrant implementation in a database. Pfaltz (1990) modernizes the point made by Aldo Leopold that, regardless of the rate of technological advancement, our ability to collect data will exceed our ability to maintain it in databases. If data are from a specific set of experimental manipulations linked only loosely to any particular place and time, they may have little value beyond use in a publication describing the result of the experiment. Similarly, data collected using non-standard methodologies may be difficult or impossible to utilize in syntheses. This is not to say that all experimental data or those data collected using non-standard methodologies should not be stored in a database. There are many examples of experimental evidence that has to be reinterpreted, and access to the data may be critical to that process. Similarly, data collected using non-standard methodologies can be integrated with those collected by other methods if a reasonable degree of caution is exercised. However, if a clear scientific need cannot be identified for a given database, it may not be reasonable to devote resources to that database.

The second question that needs to be asked is 'Who will be the *users* of the database?' This question is important on two levels. First, if a community of users for the database cannot be identified, the need for the database should be re-examined. Second, defining the users of the database provides guidance on what database capabilities will be critical to its success. For example, a database designed for use by experts in a given field is likely to be too complicated for use by elementary school students. Ideally, a database should provide data to users in a way that maximizes their immediate utility. Data need to be made available in a form such that users can manipulate them. A Hypertext Mark-up Language (HTML) table may provide an attractive way for viewing data on a WWW browser, but it can be difficult to extract data from that table in a spreadsheet or statistical package. The technological infrastructure needed to use and interpret data should be available (and preferably in common use) by the users or there is a risk that the database won't be used (Star & Ruhleder 1996).

The third question is 'What types of *questions* should the database be able to answer?' The answer to this does much to dictate how data should be structured within the database. The data can be structured in a way that maximizes the efficiency of the system for the most common types of queries, or it may be reasonable to provide multiple indices or representations of data that are applicable to different questions. For example, a large bibliographic database needs to support searches based on author, keywords and title but probably doesn't need to support searches based on page number.

The final question is one that is often not asked, but the answer to which has much to do with the success of a database: 'What *incentives* will be available for data providers?' Any database is dependent upon one or more sources of data. The current scientific environment provides few rewards for individuals who contribute data to databases (Porter & Callahan 1994). It is no accident that areas where databases have been particularly successful (e.g. genome databases) are those where contribution of data to databases is an integral part of the publication process. Leading genomic journals do not accept articles if the data have not been submitted to one of several recognized sequence databases. In the absence of support from the larger scientific community, databases need to be innovative in giving value back to the data provider. This return can be in the form of improved error checking, manipulation into new, easier-to-use forms and improvements in data input and display (see Chapter 6). Assuring proper credit to those who collected the data is a critical incentive for data contribution. Originators (authors) of data sets are more likely to make future contributions if previous contributions are acknowledged. Ideally this credit should be in a formal citation (rather than an informal acknowledgement) that specifies the originator, date, data set title and 'publication' information.

In making the myriad decisions needed to manage a database, a clear set of priorities is the developer's most valuable friend. Every database has some things it does well and some areas needing improvement. The process of database evolution is cyclical. A database may be implemented using state-of-the-art software, but several years later the state-of-the-art has advanced and the system needs to 'migrate' to new software. Therefore, database systems should be based on current priorities but have a migration path to future systems. When making decisions about the types of software to use in implementing the database and associated interfaces, it is critical to consider an 'exit strategy'. Software that stores data in proprietary formats and provides no 'export' capabilities are to be avoided at all costs.

The need for foresight applies to more than just software. The priorities of users may change. A keyword search capability may be a top user priority, but a spatial search capability may be increasingly important once it exists. It is

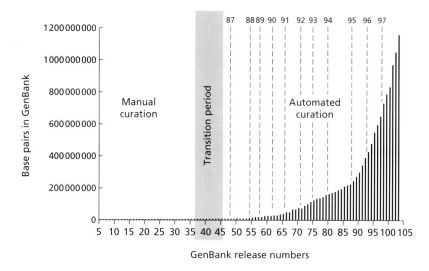

Fig. 3.2 Size of the GenBank database over time. (Adapted from a 1997 CODATA presentation, by permission of Robert J. Robbins.)

not usually possible to implement a database system *in toto*. The strategy adopted for development must recognize that, although some capabilities are not currently implemented, the groundwork for those capabilities in future versions must be provided. Thus, although an initial system may not support spatial searching, collecting and storing spatial metadata in a structured (i.e. machine-readable) form is highly desirable.

An important form of foresight is seeking scaleable solutions. Scalability means that adding or accessing the 1000th piece of data should be as easy (or easier) as adding the first. The genome databases faced a crisis when the flow of incoming data began swamping the system, which depended on some level of manual management of inputs (Fig. 3.2). The subsequent adoption of completely automated techniques for submission and quality control allowed the genome databases to handle the ever-increasing flow of data. Every system has some bottlenecks; identifying and eliminating these before they become critical is the hallmark of good planning and management.

3.4 Types of database systems

Database systems have inherent structure that reflects the basic interworkings and relationships of the database. This structure is either part of the database management system (DBMS) software or is defined within the code of more homegrown systems. Today, most systems use a relational database structure but others exist and may be well suited to particular types of data.

Table 3.1 Database system types and characteristics.

Type	Characteristics
File-system-based Use files and directories to organize information. Examples: Gopher information servers (not typically considered a DBMS)	Simple—can use generalized software (word processors, file managers) Inefficient—as number of files increase within a directory, search speed decreases Few capabilities—no sorting or query capabilities aside from sorting file names
Hierarchical Store data in a hierarchical system. Examples: IBM IMS database software, phylogenetic trees, satellite images in Hierarchical Data Format (HDF)	Efficient storage for data that have a clear hierarchy Tools that store data in hierarchically organized files are commonly used for image data Relatively rigid, requires a detailed planning process
Network Store data in interconnected units with few constraints on the type and number of connections. Example: Cullinet IDMS/R software, airline reservation databases	Fewer constraints than hierarchical databases Links defined as part of the database structure Networks can become chaotic unless planned carefully
Relational Store data in tables that can be linked by key fields. Examples: Structured Query Language (SQL) databases such as Oracle, Sybase and SQLserver, PC databases such as DBASE and FoxBase	Widely-used, mature technology Efficient query engines Standardized interfaces (i.e., SQL) Restricted range of data structures, may not handle image or expansive text well (although some databases allow extensions)
Object-oriented Store data in objects each of which contains a defined set of methods for accessing and manipulating the data. Examples: POSTGRES database	New, developing technology Wide range of structures is extensible to handle many different types of objects Not as efficient as relational DBMS for query

Here, we take a look at some of the variety of database systems that exists (Table 3.1).

A file-system-based database, as described in Chapter 2, would typically not be considered a database system because there is no 'buffer' between the physical representation of the data (in files and directories) and applications using the data (Fig. 3.3). It lacks most of the functions commonly associated with DBMS, such as support for complex relationships among data types and files, enforcement of security and integrity and error recovery. File-system-based databases typically have a heavy reliance on operating system capabilities and independent software tools to provide at least some DBMS

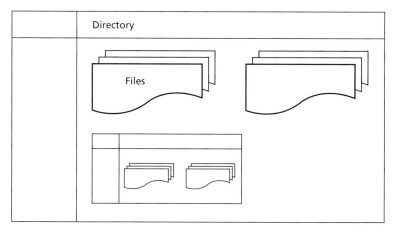

Fig. 3.3 File-system-based database.

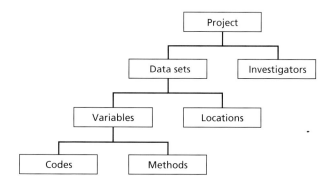

Fig. 3.4 Hierarchical database structure.

features. However, they have the advantage of relative simplicity and can be quite useful for data that do not encompass complex inter-relationships.

3.4.1 Hierarchical databases

Hierarchical databases, such as exist in the IBM IMS database software, have a higher, albeit restricted, range of structures (Hogan 1990). Data are arranged in a hierarchy that makes for efficient searching and physical access (Fig. 3.4). Each entity is linked into the hierarchy so that it is linked to one, and only one, higher-level (parent) entity, although it may be linked to multiple lower-level entities (children). Note that these relationships are defined in the design of the database and are not a function of specific data stored in the database. These designs are common in systematic databases that describe phylogenetic relationships. An example of this is the developing Universal Virus Database (Büchen-Osmond & Dallwitz 1996).

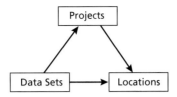

Fig. 3.5 Network database structure.

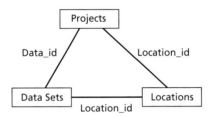

Fig. 3.6 Relational database structure.

3.4.2 Network databases

Network databases are not common in scientific applications, although they permit a wider array of relationships than hierarchical databases. Entities no longer need to be hierarchical in form (although they can be). Thus, as shown in Fig. 3.5, both projects and data sets may have links to specific locations. However, like hierarchical databases, the relationships are defined using pointers, not by the contents of the data. Modifications of relationships require physical changes be made to update pointers.

3.4.3 Relational databases

By far the most widely used database models are relational databases, which are also widely used for scientific databases. This is probably because of the relational design of the most popular DBMS software. A relational database can take on structures similar to those used in hierarchical and network databases, but with an important difference. The relational model allows inter-relationships to be specified based on key values of the data themselves. This makes it much easier to revise the structure of relational databases to add new relationships and does much to explain their popularity. In addition, relational databases benefit from a rigorous basis in mathematical theory (Bobak 1997). In Fig. 3.6, the field Data_id is shared by both the Projects and Data Sets tables and is the key field for linking those tables. Similarly, Location_id is used to link Data Sets and Projects to specific Locations. Unlike the network database (Fig. 3.5), the links in the relational database are based on the values of key fields, not explicit pointers that are external to the records themselves.

3.4.4 Object-oriented databases

Object-oriented database models are becoming increasingly common, although most frequently these models are implemented using existing relational database software to create the structure for storing object information. Query languages for object-oriented databases are still being developed and are not standard across database vendors, unlike relational databases where variations on the Structured Query Language (SQL) standard are widely used (Keller *et al.* 1998). A major feature of most object-oriented databases is the ability to extend the range of data types that can be used for fields to include complex structures (e.g. image data types). They are most frequently used with object-oriented languages such as C++ and JAVA to provide persistence to program objects (Bobak 1997). This is an area of rapid innovation (Loomis & Chaudhri 1998) that holds a great degree of promise for ecological applications (Stafford *et al.* 1994). Regardless of the type of structure, databases are usually implemented via the functionality provided by database management system (DBMS) software.

3.5 Database management system software considerations

As described in Chapter 2, the choice of software for implementation of a database must be based on an understanding of the tasks the software is expected to accomplish (e.g. input, query, sorting and analysis). Simplicity is the watchword. The software marketplace provides an abundance of sophisticated software that ranges from expensive and difficult to operate to simpler and less expensive shareware. Sophistication and complexity do not always translate to utility. The factors to be considered in choosing software extend beyond the operation of the software itself. For example, is the software in the public domain (free) or commercial? Source code for public domain software is frequently available, allowing on-site customization and debugging. An additional advantage is that file formats are usually well specified (or at least decipherable using the source code). A downside is that when something is free, sometimes it is worth every cent. Difficulty of installation, insufficient documentation, bugs in the code or lack of needed features are common complaints. In contrast, commercial software comes with technical support (often for an additional charge), is generally well documented, and is relatively easy to install. However, as for public domain software, bugs in the software are not unknown. An additional problem with commercial software is that, to some degree, you are at the mercy of the developer. Source code is almost always proprietary, and file formats frequently are proprietary as well. This can create some real problems for long-term archival storage if a commercial product is discontinued.

One consideration applicable to both public domain and commercial software is market share. Software that has a large number of users has a number of advantages over less frequently used software, regardless of specific features. A large user base provides more opportunities for testing the software. Rare or unusual bugs are more likely to be uncovered if the software is widely used. Additionally, successful software tends to generate its own momentum—spawning tools that improve the utility of the software. It is also easier to find workers with expertise in a particular software package if it is widely available.

There are numerous advantages to using DBMS software. The first is that DBMS software has many useful built-in capabilities such as sorting, indexing and query functions (Maroses & Weiss 1982; Hogan 1990). Additionally, large relational databases include extensive integrity and redundancy checks and support transaction processing with 'rollback' capabilities, allowing you to recreate the database as it existed at a particular time. There has been substantial research into making a relational DBMS as efficient as possible, and many DBMSs can operate either independently or as part of a distributed network. Finally, most DBMSs include interfaces allowing linkage to user-written programs or other software, such as statistical packages. This is useful because it allows the underlying structure of the data to be changed without having to alter programs that use the data.

Despite these advantages, most DBMSs are designed to meet the needs of business applications and these may be quite different from the needs of scientists (Maroses & Weiss 1982; Pfaltz 1990; Brunt 1994). Some functions, such as highly optimized updating capabilities, are not frequently used for scientific data because, barring detection of an error, data are seldom changed once in the database. An important disadvantage of DBMSs is that they require expertise and resources to administer. Even if a DBMS is not used for data, consideration should be given to using a DBMS for metadata (documentation). The structure of metadata conforms better to the model of business data (relatively few types of data, standard reports are useful). Most data are located based on searching metadata rather than the data themselves so the query capabilities of a DBMS are useful. This, along with the advanced interface capabilities available for DBMS make it a useful tool for the ecological database developer.

3.6 Interacting with the World Wide Web

An important innovation that applies to all DBMS software has been the introduction of software enabling DBMS to interact with World Wide Web (WWW) information servers. This makes possible a whole range of dynamic WWW pages. These pages, called web applications, can be used for both display and input, allowing users on the WWW to contribute data and metadata through a common and familiar user interface. For example, if an existing data set entry is to be edited, an interface between the DBMS (e.g. Mini-SQL, Jepson & Hughes 1998) and the WWW server allows a form to be preloaded with the existing information. When the form is submitted, changes can be made in the database immediately, so users will have immediate access to the updated information. Although a variety of proprietary user interface options for DBMS software exist, it is hard to argue against using an interface based on WWW tools. Most potential users of a database will already have access to and be familiar with a Web browser (e.g. Netscape Navigator and Microsoft Internet Explorer), so there is no need to distribute specialized software and the need for training is reduced. WWW tools continue to improve at a rapid pace. Important innovations have been the support of online forms and linking WWW servers to database engines. The addition of programming languages (such as Java) that allow secure operation of applications on the client-side have greatly increased the types of operations supported over the WWW. WWW tools can be used for input to a database as well as for output. An advantage of this approach is that input of metadata and data can be made from many different locations, which can circumvent personnel and equipment bottlenecks.

The WWW is an excellent medium for retrieving data from a DBMS for display, be it tables of numbers, graphics, audio or video. However, if you want to go beyond display, to actually transferring data directly to analytical software on a user's computer, there is much room for improvement. Current browser software must be customized to relate specific file types with specific applications. This area is rapidly evolving, though most improvements in interfacing DBMS to other applications are expected via network middleware like Microsoft's Object Data Base Connectivity (ODBC) that is independent of a browser.

3.7 Choosing a computer system

Although the selection of a computer system to support database operations is an important decision, it is no longer a dominant one. The focus has shifted from specific computer hardware to general decisions regarding the software and the types of operating systems that support them (Porter & Kennedy 1992). Rapid innovations in microcomputer technology have blurred the lines among the capabilities of mainframe computers, minicomputers and microcomputers. The choice of a system may rest more on the types of support available in the local computing environment than on specifics of the hardware or operating system capabilities.

Important considerations include sufficient computational resources to support the desired software, the capacity of the system to support a database of the size envisioned, scalability should the database size increase, ease of administration, reliability and availability of repair services. Currently, there are two reasonable options for a computer system to support database creation: computers running flavours of UNIX (Solaris, Irix, OpenBSD, Linux) and computers running Microsoft Windows NT. UNIX is a mature, full-functioned operating system. It has strong capabilities for multitasking and multiuser support. As a mature system, it is reliable and robust and many WWW tools are available, often free of cost. On the downside, UNIX is difficult to learn and commercial software for UNIX is typically much more expensive than that for personal computer-based systems.

Microsoft Windows NT is a rapidly evolving operating system. Compared to UNIX, software and hardware are relatively inexpensive and most software is more user-friendly than UNIX, though no less sophisticated. The capabilities of these systems are similar enough that choice of a system may depend on the local computational environment. If UNIX computers are already in place and there is sufficient expertise to support them, UNIX is likely the best choice. However, if those prerequisites are lacking, an NT system may be the better choice.

3.8 Interlinking information resources

Maximizing the utility of database resources requires going beyond the simple creation of individual databases. Integrative research approaches require the combination of data, often from diverse sources. Users benefit from being able to search multiple databases via a single query. (Z39.50 and CORBA are examples of international standards designed to facilitate both the client and server side of these types of queries.) In addition, the value of the data contained in an individual database is elevated when users are able to easily locate ancillary and related data found elsewhere. A frequent phenomenon accompanying development of successful databases is that they spawn a series of value added databases that tailor the raw information contained in one or more 'basic' databases to meet the needs of a specific community.

3.9 Data modelling and normalization

In the creation of databases, the DBMS constitutes the canvas, but the data model is the painting. The purpose of a data model is to explicitly define the entities represented in a database and spell out the relationships among these different entities (Bobak 1997). Ultimately, the data model will be used as the road map for the definition of tables, objects and relations. Typically, it is at a level of abstraction that lets us get past a mass of detail to look at the 'big picture'.

Table 3.2 'Flat file' species observation database.

Genus	Species	Common name	Observer	Date
Quercus	alba	White Oak	Jones, D.	15-Jun-1998
Quercus	alba	White Oak	Smith, D.	12-Jul-1935
Quercus	alba	White Oat	Doe, J.	15-Sep-1920
Quercus	rubra	Red Oak	Fisher, K.	15-Jun-1998
Quercus	rubra	Red Oak	James, J.	15-Sep-1920

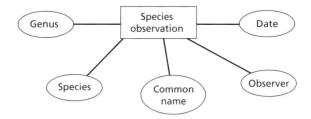

Fig. 3.7 Entity–relationship diagram for data set in Table 3.2.

Normalization is a process wherein a data model is reduced to its essential elements (Hogan 1990). The aim of normalization is to eliminate redundancies and potential sources of inconsistency. During the normalization process, it is not unusual to define new entities and attributes or to eliminate old ones from a data model. Note that data modelling is in many ways a creative process. Although there are rules for normalization, the data model inevitably reflects the purpose of the database and the thought process of its creator.

The data modelling process is best described through an example. The simplest model would be to store all data in a single table; Table 3.2 is an example of a database of species observations. Figure 3.7 shows how this table would be represented in an entity–relationship diagram (E–R diagram). The box represents a single table per entity. The ovals represent attributes (fields) within that table. By the rules of normalization, there are several deficiencies in this model. First, the table is full of redundancies. The species *Quercus alba* is represented numerous times within the table, as is the common name 'white oak'. This gives many opportunities for errors to enter the table, for example, in the third line, white oak is misspelled 'white oat'. A second option is to split the table into two entities, one representing the species-level data and another entity for the observations (Tables 3.3 & 3.4). An attribute needs to be added that can link the two entities together (here called Spec_code). This key attribute can be any unique code, including numbers; here the codes are based on the genus and species names. Figure 3.8 is a schematic illustrating the link (dotted line) between entities provided by this attribute.

Table 3.3 Table for species entity.

Spec_code	Genus	Species	Common name
QRCALB	Quercus	alba	White Oak
QRCRBR	Quercus	rubra	Red Oak

Table 3.4 Data table for observations entity.

Spec_code	Observer	Date
QRCALB	Jones, D.	15-Jun-1998
QRCALB	Smith, D.	12-Jul-1935
QRCALB	Doe, J.	15-Sep-1920
QRCRBR	Fisher, K.	15-Jun-1998
QRCRBR	James, J.	15-Sep-1920

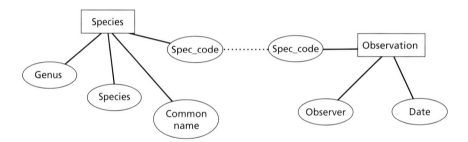

Fig. 3.8 Example data model.

In a real application, the species entity would incorporate all data relevant to the species that are independent of any specific observation. Additional attributes of the species entity might include additional taxonomic information (e.g. family, order), physical characteristics (e.g. mean size, branching pattern, leaf type) and natural history information (e.g. reproductive characteristics, habitat). It may be desirable to include images or Internet links. The observation entity might be expanded to include information on the observation (e.g. method of reporting, citation, voucher specimens), on the location of the observation and additional details on the observer (e.g. contact information, such as address and e-mail). As this process proceeds it may become evident that additional entities are required. For example, if a single observer makes multiple observations, it may make sense to establish an observer entity with attributes such as address, phone number, e-mail, along with a new key attribute: Observer_code. Similarly, we might want to add an additional entity describing locations, including coordinates, habitats, etc. Linking

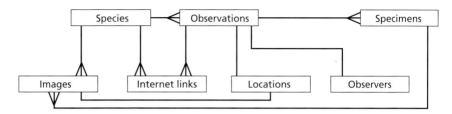

Fig. 3.9 Data model for species observation data set.

images to specimens and locations as well as species could further extend the model. A more comprehensive (but by no means exhaustive) data model showing the entities (but not attributes) for a species observation database is shown in Fig. 3.9. Note that multiple lines connecting the entities depict a one-to-many relationship. Thus, a species may have multiple observations, Internet links and images, but each observation, Internet link or image may be linked to only one species. As you can see, the process can quickly become complex.

A number of software tools are now available to aid the developer in the data modelling process. These tools allow you to maintain multiple logical and physical views of a database, to validate and test the data models, to reverse engineer a database from one platform and implement it on another, and have a variety of tools for publishing design and data dictionary information. Two of the more common ones now in use are ERWin and ER/Studio.

3.10 Examples of scientific databases

Scientific databases can be lumped into one of two categories based on content: deep or wide. 'Deep' databases specialize in a single or a few types of data and implement sophisticated searching and analytical capabilities (Table 3.5). Examples of this type of database are Genbank, which serves as a primary archive of genetic sequence data for the human genome project, with over

Table 3.5 'Deep' vs. 'wide' databases.

'Deep' databases	'Wide' databases
• Specialize on one or a few types of data	• Contain many different kinds of data
• Large numbers of observations of one (or few) type(s) of data	• Many different kinds of observations, but relatively few of each type
• Provide sophisticated data query and analysis tools	• May provide tools for locating data, but typically do not have tools for analysis
• Tools operate primarily on data content	• Tools operate primarily on metadata content

one billion bases in approximately 1.6 million sequences (National Center for Biotechnology Information 1997); SWISS-PROT, which contains 69 113 protein sequence entries, abstracted from 59 101 references (Bairoch 1997); and PDB, the protein structure database, which contains over 6000 atomic coordinate entries for protein structure (Protein Databank 1997). These are very large databases with funding in excess of one million US dollars per year.

There is also a variety of specialized databases operating on a smaller scale. For example, MUSE is specialized software for managing herbarium specimens (Humphries 1997) and BIOTA is software for management of specimen-based biodiversity data (Colwell 1997). These systems are commercially available and are used by a variety of institutions and investigators.

'Wide' databases are data collections that attempt to capture all the data related to a specific field of science (Table 3.5). For example, the US National Geophysical Data Center (NGDC), operated by the US National Oceanic and Atmospheric Administration (NOAA), supports over 300 databases containing geophysical data (NGDC 1997). Such data centers use standardized forms of metadata (e.g. GILS, FGDC, DIF) for maintaining formal catalogues with controlled vocabularies for subjects and keywords (see Chapter 5). Similarly, the US National Aeronautic and Space Administration (NASA) operates a series of Distributed Active Archive Centers (DAACs), each of which specializes in supporting a particular area of earth or space science and has a varying number of different types of data sets (see Chapter 6).

Other databases that can be classified as wide are project-based databases. These databases support a particular multi-disciplinary research project and may include a wide array of data focused on a particular site or research question. Examples of this type of database are those at Long Term Ecological Research (LTER) sites. These databases contain a wide array of ecological data (e.g. weather and climate, primary productivity, nutrient movements, organic matter, trophic structure, biodiversity and disturbance), along with information that supports management of the site (e.g. researcher directories, bibliographies and proposal texts).

Some databases, such as those created by individual researchers may not fit the classification of 'wide' or 'deep'. The level of development of such databases varies tremendously, as does the quality and quantity of the associated metadata. As an aggregate, these databases constitute a valuable resource, but one that is difficult to exploit because data can be hard to locate and metadata may be insufficient or difficult to translate into usable forms. Brackett (1996) described such 'massively disparate' data as 'locally useful and globally junk'.

No database exists in isolation from its user community. In order to satisfy that community, a database needs to meet a variety of requirements.

1 It must contain data that are wanted by a group of users.

2 Data must be current and as complete as possible. A corollary of this requirement is that data must be easily added. A second corollary is that there must be incentives for data providers.
3 The database must provide the data to users in a way that is attractive to them.
4 The level of technological sophistication needed to use the system must match that of the users.
5 There needs to be a mechanism for dealing with user queries and questions and for identifying unmet user needs. To be successful, a database needs to meet *all* of these requirements.

3.10.1 Databases related to the Human Genome Project

The Human Genome Project provides an excellent case study of the opportunities and pitfalls inherent in linking databases together. The data from GenBank, EMBL and the Genome Data Base serve as 'grist for the mill' of other databases. For example, the Ribosomal Database Project (Maidak *et al.* 1997) harvests data on RNA data sequences from releases of GenBank. RDP then performs additional analyses to align sequences from different sources and develop phylogenetic diagrams. It also provides specialized tools for locating 'probes' that may be used to distinguish classes of sequences. The RDP is then used by communities of ecological and health researchers to identify microbes or, in the case of unknown microbes, to estimate the probable characteristics of such microbes based on their similarity to known microbes. Although it contains no raw data that are not available in sequence databases, RDP and similar databases reduce the duplication inherent in having each individual researcher analyse the raw data.

3.10.2 Databases related to worm ontogeny

At the time it was created (pre-WWW), the Worm Community System was a technological *tour de force*. This specialized database, produced by Bruce Schatz (Schatz 1992), integrated many different kinds of data on the worm *Caenorhabditis elegans*. This worm is unique in that its entire embryological development has been mapped, allowing creation of a 'wiring diagram' relating any given cell to its progenitor in the developmental process. There are also extensive genetic data. These data sources, along with hypertext versions of published literature, were linked together with a single user interface that allowed point-and-click access to these data. Specialized tools for display were developed. For example, a display tool for embryological development created an interactive tree-diagram that could be zoomed to look at the level of the individual cell. Importantly, the different types of data were linked together. If a paper mentioned a given gene or cell, a link in the hypertext version of the paper would provide immediate access to information on that gene or cell. Once a cell or gene was selected, all papers relating to that gene or cell could be called to the screen. The database had strong links to the community of *C. elegans* researchers, who committed to providing data.

Despite these state-of-the-art features (or perhaps because of them) the Worm Community System is less widely used than other much less sophisticated databases of *C. elegans*. Star and Ruhleder (1996) examined the successes and failures of the Worm Community System in relation to the community of *C. elegans* researchers. They found a critical factor that greatly diminished the use of the system was the technological gap between the users and the system. Use of the Worm Community System required a UNIX computer running the X-Windows windowing system. However, most of the *C. elegans* researchers used PC or Macintosh computers. Many laboratories were unable to justify the expenditure for a new, expensive computer system, especially one where there was no local body of expertise to administer it. Additionally, generating the complex of hypertext links for new data entering the system meant it was slower making information available than simpler systems that did not provide linkages.

In some ways, the Worm Community System was simply ahead of its time. The advent of WWW browsers now virtually erases the technological gap between a system of comparable complexity and its users. Had the WWW been in wide use, the gap between the technology levels of users and the system would have been greatly reduced. There now exists a plethora of tools for easily creating HTML documents. However, there would still remain issues regarding addition of data and the generation of linkages between different types of data.

3.11 Discussion

Scientific databases are increasingly setting the boundaries for science itself. Subsequently, taking ecological science to the next step will require taking ecological databases to a new level. A key to the success of scientific databases lies in developing incentives for the individuals who collect data to make those data available in databases (Porter & Callahan 1994). Additionally, mechanisms for funding databases need to be developed. For-profit databases have been successful in some areas (e.g. Chemical Abstracts) but are an unlikely candidate for success where the number of potential users is small (regardless of the importance of the data to our understanding of nature). Direct funding of databases has had some successes, but in many ways a successful database is a funding agency's worst nightmare: a project that grows year after year and never goes away. This model has the additional deficiency in that it decouples database funding from database use. Ideally, a database should have an organic link with its user community, and top-down funding can dissipate that link. The correct model probably lies between for-profit and agency-funded databases. Robert Robbins (1997) proposes an innovative scheme where funding agencies give each grantee a set of 'data chits' to be spent on information management services, thus providing both additional funding to support databases and maintaining the link between a database and its user community.

Trends in database and computational technology hold great promise for the future development of scientific databases. The move of businesses towards large-scale integrated data resources known as data warehouses (Brackett 1996) vastly increases the resources applied to dealing with problems of diversity. We have already seen a huge impact by the advent of the WWW. However, network technologies are still in their infancy. Increased network speed (bandwidth), improved server-side and client-side programs, wireless communications and improved software links to databases can significantly ease the difficulty of creating scientific databases. Extensions to relational databases and the continuing evolution of object-oriented databases will increase the ease of dealing with diverse types of data, such as images, audio and video (Stafford *et al.* 1994).

Scientific databases evolve; they don't spontaneously generate. We are at an exciting time in the development of scientific databases. Scientific questions and technological advances are coming together to make a revolution in the availability and usability of scientific data possible. However, the ultimate success of scientific databases will depend on the commitment of individuals and organizations starting and operating databases.

3.12 References

Bairoch, A. (1997) *The SWISS-PROT Protein Sequence Data Bank User Manual.* Release 35. (http://www.expasy.ch/txt/userman.txt)

Bobak, A. (1997) *Data Modeling and Design for Today's Architectures.* Artech House, Boston, MA.

Büchen-Osmond, C. & Dallwitz, M.J. (1996) Towards a universal virus database—progress in the ICTVdB. *Archive of Virology,* **141**, 392–399.

Brackett, M.H. (1996) *The Data Warehouse Challenge: Taming Data Chaos.* John Wiley & Sons, New York.

Brunt, J.W. (1994) Research data management in ecology: a practical approach for long-term projects. In: *Seventh International Working Conference on Scientific and Statistical Database Management.* (eds J.C. French & H. Hinterberger), pp. 272–275. IEEE Computer Society Press, Washington, DC.

Cinkosky, M.J., Fickett, J.W., Gilna, P. & Burks, C. (1991) Electronic data publishing and GenBank. *Science* **252**, 1273–1277.

Colwell, R.K. (1997) *Biota: The Biodiversity Database Manager.* (http://viceroy.eeb.uconn.edu/biota)

Gilbert, W. (1991) Towards a paradigm shift in biology. *Nature* **349**, 99.

FGDC. (1994) *Content Standards for Digital Spatial Metadata (June 8 Draft).* Federal Geographic

Data Committee. Washington, D.C. (http://geochange.er.usgs.gov/pub/tools/metadata/standard/metadata.html)
Hogan, R. (1990) *A Practical Guide to Data Base Design*. Prentice Hall, Englewood Cliffs, NJ.
Humphries, J. (1997) *MUSE*. (http://www.keil.ukans.edu/muse/)
Jepson, B. & Hughes, D.J. (1998) *Official Guide to Mini SQL 2.0*. John Wiley & Sons, New York.
Justice, C.O., Bailey, G.B., Maiden, M.E., Rasool, S.I., Strebel, D.E. & Tarpley, J.D. (1995) Recent data and information system initiatives for remotely sensed measurements of the land surface. *Remote Sensing and the Environment* **51**, 235–244.
Keller, W., Mitterbauer, C. & Wagner, K. (1998) Object-oriented data integration: running several generations of database technology in parallel. In: *Object Databases in Practice*. (eds M.E.S. Loomis & A.B. Chaudhri), pp. 3–10. Prentice Hall, Upper Saddle River, NJ.
Loomis, M.E.S. & Chaudhri, A.B. (1998) *Object Databases in Practice*. Prentice Hall, Upper Saddle River, NJ.
Maidak, B.L., Olsen, G.J., Larsen, N., Overbeek, R., McCaughey, M.J. & Woese, C.R. (1997) The RDP (Ribosomal Database Project). *Nucleic Acids Research* **25**, 109–111.
Magnuson, J.J. (1990) Long-term ecological research and the invisible present. *BioScience* **40**, 495–501.
Maroses, M. & Weiss, S. (1982) Computer and software systems. In: *Data Management at Biological Field Stations*. (eds G. Lauff & J. Gorentz), pp. 23–30. A report to the National Science Foundation. W.K. Kellogg Biological Station, Michigan State University, Hickory Corners, MI.
Meeson, B.W. & Strebel, D.E. (1998) The publication analogy: A conceptual framework for scientific information systems. *Remote Sensing Reviews* **16**, 255–292.
Michener, W.K., Brunt, J.W., Helly, J., Kirchner, T.B. & Stafford, S.G. (1997) Non-geospatial metadata for the ecological sciences. *Ecological Applications* **7**, 330–342.
National Center for Biotechnology Information. (1997) *GenBank Overview*. (http://www.ncbi.nlm.nih.gov/Genbank/GenbankOverview.html)
NGDC. (1997) National Geophysical Data Center. (http://www.ngdc.noaa.gov/)
NRC. (1997) *Bits of Power: Issues in Global Access to Scientific Data*. (http://www.nap.edu/readingroom/books/BitsOfPower/) National Research Council. National Academy Press, Orlando, FL.
Pfaltz, J. (1990) Differences between commercial and scientific data. In: *Scientific Database Management*. (eds J.C. French, A.K. Jones & J.L. Pfaltz), Technical Report 90–22, pp. 125–129. Department of Computer Science, University of Virginia, Charlottesville, VA. (http://www.lternet.edu/documents/Reports/Data-and-information-management/UVA_CS_90/cs_90-22.NR)
Porter, J.H. & Callahan, J.T. (1994) Circumventing a dilemma: historical approaches to data sharing in ecological research. In: *Environmental Information Management and Analysis: Ecosystem to Global Scales*. (eds W.K. Michener, J.W. Brunt & S.G. Stafford), pp. 193–203. Taylor and Francis, Ltd., London.
Porter, J.H. & Kennedy, J. (1992) Computer systems for data management. In: *Data Management at Biological Field Stations*. (ed. J.B. Gorentz), pp. 19–28. A Report to the National Science Foundation. W.K. Kellogg Biological Station, Michigan State University, Hickory Corners, MI.
Protein Databank. (1997) PDB Newsletter, October 1997. PDB release 82. Brookhaven National Laboratory, Upton NY. (ftp://ftp.rcsb.org/pub/pdb/doc/newsletters/bnl/newsletter97oct/)
Robbins, R.J. (1994) Biological databases: A new scientific literature. *Publishing Research Quarterly* **10**, 1–27.
Robbins, R.J. (1995) Information infrastructure. *IEEE Engineering in Medicine and Biology* **14**, (6), 746–759.
Robbins, R.J. (1997) *Next steps for working scientists: access to data*. Presentation to CODATA Conference on Scientific and Technical Data Exchange and Integration. National

Institutes of Health, Bethesda, Maryland, December 1997. (http://www.esp.org/rjr/codata.pdf)

Schatz, B.R. (1992) Building an electronic community system. *Journal of Management Information Systems* **8**, (3), 87–101. (http://www.canis.uiuc.edu/publications/)

Stafford, S.G., Brunt, J.W. & Michener, W.K. (1994) Integration of scientific information management and environmental research. In: *Environmental Information Management and Analysis: Ecosystem to Global Scales.* (eds W.K. Michener, J.W. Brunt & S.G. Stafford), pp. 3–19. Taylor and Francis, Ltd., London.

Star, S.L. & Ruhleder, K. (1996) Steps toward an ecology of infrastructure: design and access for large information spaces. *Information Systems Research* **7**, 111–134.

Strebel, D.E., Meeson, B.W. & Nelson, A.K. (1994) Scientific information systems: a conceptual framework. In: *Environmental Information Management and Analysis: Ecosystem to Global Scales.* (eds W.K. Michener, J.W. Brunt & S.G. Stafford), pp. 59–85. Taylor and Francis, Ltd., London.

Strebel, D.E., Landis, D.E., Huemmrich, K.F., Newcomer, J.A. & Meeson, B.W. (1998) The FIFE data publication experiment. *Journal of Atmospheric Sciences* **55**, 1277–1282.

CHAPTER 4

Data Quality Assurance

DON EDWARDS

4.1 Introduction

The importance of data quality to ecological research cannot be overstated, yet the topic of quality assurance receives surprisingly little attention in the scientific literature. This chapter cannot address all facets of this topic. It will touch on several aspects integral to any data quality assurance programme and, in particular, will focus on the detection of outliers in data as an intermediate step in eliminating data contamination.

4.2 Prevention first

Data contamination occurs when a process or phenomenon other than the one of interest affects a variable value. Prevention of data contamination through quality control is by far preferable to after-the-fact heroics. American industry learned the prevention lesson the hard way in the 1960s and 1970s, when advancements in quality science in Japan erased the worldwide dominance of United States companies in the electronics and automobile industries. Ironically, Americans Joseph Juran and W. Edwards Deming, sent to Japan after World War II to help with reconstruction, played huge roles in this Japanese 'coup'. The relative importance of prevention was expressed succinctly by the ever-acidic Deming in his 1986 book *Out of the Crisis*: 'Let's make toast the American industry way—you burn, I'll scrape'.

Prevention is primarily a data management issue, not a statistical one. In theory, many management strategies designed for manufacturing and service-industry quality assurance could be useful for data quality assurance in large-scale scientific endeavours. Many of the quality problems encountered in manufacturing or providing a service also occur in the process of database construction and management. For example, Flournoy and Hearne (1990), in the setting of a cancer research centre, stress the importance of all users of, and data contributors to, a multi-user database having a stake in data quality. In fact, this is also one of Deming's (1986) principles: all company employees, from upper-level management (principal investigators) to line-workers (data entry technicians), must feel a responsibility for, and a pride in, product (database) quality. Of course, the real challenge lies in inspiring this universal motivation.

One successful management technique that could facilitate scientific data quality assurance is the use of 'quality circles'. These would be brief, regular (e.g. weekly) meetings of scientists, technicians, systems specialists and data entry personnel for discussing data quality problems and issues. Everyone, including upper-level management, should have a basic understanding of natural variability and simple statistical methods for dealing with variability (Deming 1986). These meetings would build teamwork attitudes while focusing brainpower on data quality issues. Participants become constantly aware of quality issues and learn to anticipate problems. Moreover, all personnel better appreciate the importance of their role in data quality and the entire scientific effort.

4.3 Preventing data contamination

4.3.1 Data entry errors

Sources of data contamination due to data entry errors can be eliminated or greatly reduced by using quality control techniques. One very effective strategy is to have the data keyed independently by two data entry technicians and then computer-verified for agreement. This practice is commonplace in professional data entry services and in some service industries such as the insurance industry (LePage 1990). Scientific budgets for data entry often do not include funds for double keying of data. Unfortunately, other means of detecting keypunch errors (e.g. application of after-the-fact outlier-detection tests by masters- or doctoral-level personnel) are less effective and more expensive since they involve higher-paid personnel.

4.3.2 Illegal data filters

Illegal data are variable values or combinations of values that are impossible given the actual phenomenon observed. For example, no values for a count variable (e.g. the number of flowers on a plan' outside of the interval [0,1] for a proportion variable would be Illegal combinations occur when natural relationships values are violated, for example, if Y_1 is the age of a bande' census, and Y_2 is the same bird's age in this year's censu' be less than Y_2. These kinds of illegal data often occu but may occur for other reasons, such as misreadin' of observations in the field or laboratory.

A simple and widely used technique for det' ination is an illegal data filter. This is a comput 'laundry list' of variable value constraints

Table 4.1 An illegal-data filter, written in SAS. The data set 'All' exists prior to this DATA step, containing the data to be filtered, variable names Y1, Y2, etc., and an observation identifier variable ID.

Data Checkum; Set All;
  message=repeat(" ",39);
  If Y1<0 or Y1>1 then do; message="Y1 is not on the interval [0,1]"; output; end;
  If Floor(Y2) NE Y2 then do; message="Y2 is not an integer"; output; end;
  If Y3>Y4 then do; message="Y3 is larger than Y4"; output; end;
  :
  (add as many such statements as desired . . .)
  :
  If message NE repeat(" ",39);
  keep ID message;
Proc Print Data=Checkum;

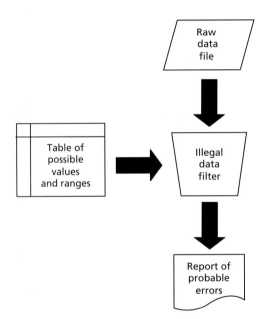

Fig. 4.1 Flow of information in the use of an illegal data filter to detect data contamination.

te to be added to the master) and creates an output data set. This output et includes an entry for each violation and identifying information and a ge explaining the violation. Table 4.1 and Fig. 4.1 show the structure of rogram, written in the SAS® language (SAS 1990). The filter program pdated and/or enhanced to detect new types of illegal data that may been anticipated early in the study.

d of caution must be interjected regarding the operative word 'illegal'. Simply because one has never observed a concentration below a given threshold, and can't imagine it ever happening, does not make such an observation an illegal data point. One of the most famous data QA/QC blunders to date occurred when the US National Aeronautics and Space Administration's computer programs deleted satellite observations of ozone concentrations that were below a specified level, which greatly delayed the discovery of the ozone hole over the South Pole (Stolarski *et al.* 1986).

4.4 Outlier detection

4.4.1 Philosophy

The term outlier is not (and should not be) formally defined. An outlier is simply an unusually extreme value for a variable, given the statistical model in use. What is meant by 'unusually extreme' is a matter of opinion, but the operative word here is 'unusual'. After all, some extreme values are to be expected in any data set. The formidable challenge of outlier detection lies in deciding how unusual an extreme point must be before it can (with confidence) be considered 'unusually' unusual. It must be emphasized, and will be demonstrated, that the outlier notion is model-specific. A particular value for a variable might be highly unusual under a linear regression model but not unusual in an analysis without the regressor. So, outlier detection is part of the process of checking the statistical model assumptions, a process that should be integral to any formal data analysis.

'Elimination of outliers' should not be a goal of data quality assurance. Many ecological phenomena naturally produce extreme values, and to eliminate these values simply because they are extreme is equivalent to pretending the phenomenon is 'well-behaved' when it is not. To mindlessly or automatically do so is to study a phenomenon other than the one of interest. The elimination of data contamination is the appropriate phrasing of this data quality assurance goal. If contamination is unnoticed at observation time, it can usually only be detected later if it produces an outlying data value. Hence, the detection of outliers is an intermediate step in the elimination of contamination. Once the outlier is detected, attempts should be made to determine if some contamination is responsible. This would be a very labour-intensive, expensive step if outliers were not, by definition, rare. Note also that the investigation of outliers can in some instances be more rewarding than the analysis of the 'clean' data: the discovery of penicillin, for example, was the result of a contaminated experiment. If no explanations for a severe outlier can be found, one approach is to formally analyse the data both with and without the outlier(s) and see if conclusions are qualitatively different.

4.4.2 Checking test assumptions with normal probability plots

Formal statistical tests to detect outliers nearly always require some pre-specification of 'tail behaviour' for the measurement of interest. In other words, we must somehow specify the typical patterns of extreme values to be expected in an uncontaminated data set of a given size. In order to do this, the test will usually assume that the uncontaminated measurements follow a given probability distribution, usually the normal (or Gaussian) distribution. To this author's knowledge, every formal outlier detection rule and/or test makes this distributional assumption or an analogous one, and will not work well if the distributional assumption is substantively violated. This is an enormous drawback for formal outlier detection tests; it is unwise to use them without thoroughly checking their distributional assumptions. This sensitivity of outlier tests does not hold for all statistical procedures that nominally assume normality; for example, t-tests and F-tests in ANOVA and regression are typically very robust to this assumption. That is, they work well for moderate-to-large samples even for substantially non-normal data. This is because they make statements comparing long-run mean values for the measurement, not statements about extreme values.

Since most outlier tests assume that the measurements of interest (the 'errors' in a regression or ANOVA model) follow a normal distribution, a few words are in order regarding methods for checking this assumption. There are many ways to do this, for example, via simple histograms or formal tests of normality such as the Kolmogorov-Smirnov test (Sokal & Rohlf 1995). An old means of checking normality that is gaining increased popularity in the computer age is the normal probability plot. Chambers *et al.* (1983) provide an excellent discussion of these and other quantile plots for checking distributional assumptions.

The idea behind the normal probability plot, often attributed to Daniel (1959), is as follows: an idealized histogram of data from a normal distribution follows the bell-shaped curve (Fig. 4.2a). If at each potential variable value, Y, we calculate the cumulative area under the bell curve to its left, say $\Phi(Y)$, and then plot these, we obtain the sigmoid curve shown in Figure 4.2b. These $\Phi(Y)$ values are in fact the values provided in a table of the normal probability distribution. A normal probability plot essentially re-spaces the vertical axis so that points following this particular sigmoid curve, when plotted against the re-spaced axis, fall on a straight line. If sample points deviate substantially from a line when plotted in this way, they come from a non-normal distribution. Moreover, the pattern of non-linearity in the plot can provide useful information as to the type of non-normality in the data, as any given test will be sensitive to certain types but possibly not to others.

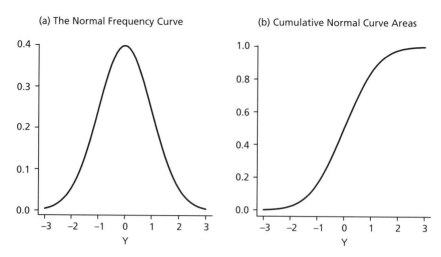

Fig. 4.2 The normal density and cumulative distribution functions.

Table 4.2 SAS and Splus commands for constructing Normal probability plots. Each code assumes the pre-existence of a data set with variable Y containing the sample values to be plotted.

SAS for a Normal probability plot:
  Proc Rank Normal=Blom out=All;
    Var Y; Ranks=Nscores;
  Proc Plot;
    Plot Y*Nscores;

Splus for a Normal probability plot:
  >qqnorm(Y)

Prior to the advent of today's computing power, researchers used special graph paper (probability paper) with the appropriate vertical axis spacing, to make normal probability plots. Today, these plots are constructed using computer software that creates the normal scores for the sample and then plots the sample values against these. If the ordered sample values are denoted $Y_1 < Y_2 < \ldots < Y_n$, the corresponding normal scores are $\Phi^{-1}[1/(n+1)]$, $\Phi^{-1}[2/(n+1)], \ldots, \Phi^{-1}[n/(n+1)]$. (Note: there have been minor variations of this formula proposed by various researchers; for example, the SAS code above uses the formula proposed by Blom 1958). The function $\Phi^{-1}(p)$, sometimes called the probability inverse or probit function, provides the pth percentile of the normal distribution for any p, $0 < p < 1$. Table 4.2 shows code in SAS and Splus® for constructing normal probability plots. Figure 4.3 shows an Splus plot of 30 sample values taken from a normal distribution.

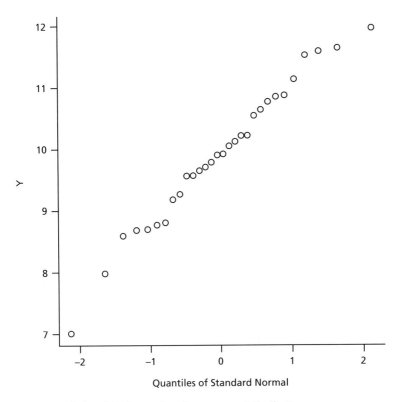

Fig. 4.3 A normal plot of 30 observations from a normal distribution.

Examples of certain patterns that may occur in the normal probability plot when the data are seriously non-normally distributed are shown in Fig. 4.4. Data from a severely right-skewed distribution (i.e. one that produces extreme large values more often than small ones) is plotted in Fig. 4.4a. Figure 4.4b shows a similar pattern, but this time with data from a left-truncated normal distribution; such data might arise from a normally distributed variable whose values frequently drop below detection limits. Clearly, the normal plot can be a useful informal means of detecting outliers as the outliers will 'fall' to the lower left or upper right of the line formed by the preponderance of the sample values (if these are normally distributed) (see Figs 4.4c & 4.4d). To reiterate, an outlier is not necessarily contamination; many real-world phenomena naturally produce extreme values more often than would be found in a normal distribution. These 'fat-tailed' distributions can result in normal plots like that of Fig. 4.4c, which is very similar to the contaminated-sample plot in Fig. 4.4d. There is no substitute for examining the history of any outlying points in order to ascertain if they are due to contamination or represent natural extremes for the phenomenon of interest.

DATA QUALITY ASSURANCE

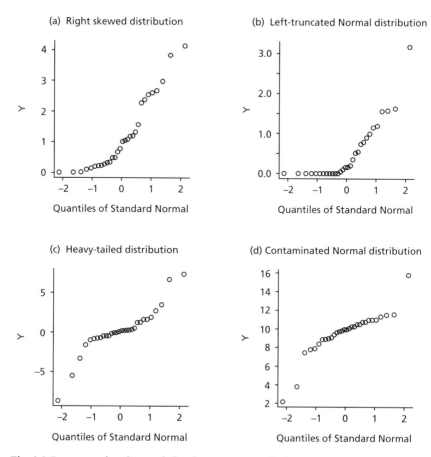

Fig. 4.4 Four examples of normal plots from non-normally distributed data.

4.5 A formal outlier test: Grubbs' test

One of the oldest and most widely used procedures for detecting contamination in samples is Grubbs' test (Grubbs & Beck 1972; ASTM E 1994). By 'samples' we mean that, if the data are uncontaminated, we would have several (say, n) independent observations on the variable from the same repeatable, well-defined, stable experimental process. Grubb's test assumes the uncontaminated process produces data that follow a normal distribution, and it is very sensitive to that assumption; if the 'clean' data are clearly non-normally distributed, one should not use Grubbs' test.

Grubbs' test is performed as follows: let $Y_1 < Y_2 < \ldots < Y_n$ again denote the ordered sample values, and $\bar{Y}$ and S the sample mean and standard deviation, respectively. If it is only of interest to detect unusually large outliers, then compare the test statistic

$$T_n = (Y_n - \bar{Y})/S$$

to the appropriate one-sided critical point (Grubbs & Beck 1972; ASTM E 1994), which increases with n, and has an error rate denoted α_G. If it is only of interest to detect unusually small outliers, compare the test statistic

$$T_1 = (\bar{Y} - Y_1)/S$$

to the appropriate one-sided critical point. If either large or small outliers are to be detected, compare the larger of T_n and T_1 to the two-sided critical point.

The probability α_G is, in this case, a per-sample error rate. Users are encouraged to choose α_G thoughtfully, as it has a different meaning than the α-level one uses in testing research hypotheses. What fraction of the clean data are you willing to lose, or at the very least investigate, for the sake of detecting possible contamination? For example, if α_G is chosen to be 0.05, then in 5% (one in 20) of repeated, uncontaminated samples of this size, we would falsely declare a contamination to exist using Grubbs' test. Bear in mind that if such contamination is really severe, it would be detected using a smaller α_G. ASTM E (1994) recommends a 'low significance level, such as 1%'. It should also be noted that Grubbs' test cannot be performed when n = 2, and the critical points for n = 3 do not differ for choices of two-sided α_G less than 0.05.

As an example of the (mis-)application of Grubbs' test, consider the seeded-cloud rainfall data of Simpson et al. (1975) shown in Table 4.3. The mean and standard deviation for these data are $\bar{Y} = 442$ and $S = 651$. With n = 26 and $\alpha_G = 0.01$, the one-sided critical point for Grubbs' test is 3.029, and the test statistic for detecting large outliers is $T_{26} = (2745.6 - 442)/651 = 3.539$. Hence (if being careless) we would assert contamination in this case.

Of course, the assumption that the uncontaminated sample follows a normal distribution is grossly violated here. Figures 4.5a and 4.5b show a histogram and normal probability plot for the raw data, which clearly show the sample, as a whole, follows a severely right-skewed distribution. Figures 4.5c and 4.5d show a histogram and normal plot for the $\log_{10}$-transformed rainfall data. Clearly, these rainfall data are very nearly log-normally distributed, and there is no evidence of contamination. This example demonstrates the importance of visual inspection of plots and histograms in the quality assurance process since total reliance on computer-generated statistical tests may be misleading.

Table 4.3 Rainfall in acre-feet from seeded clouds (Simpson et al. 1975).

4.1	7.7	17.5	31.4	32.7	40.6	92.4	115.3	118.3
119.0	129.6	198.6	200.7	242.5	255.0	274.7	274.7	302.8
334.1	430.0	489.1	703.4	978.0	1656.0	1697.8	2745.6	

DATA QUALITY ASSURANCE   79

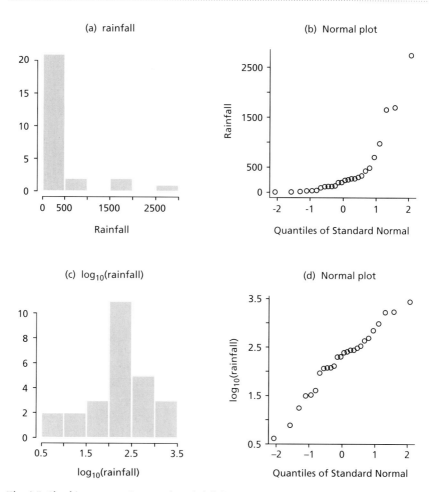

Fig. 4.5 Checking assumptions on the rainfall data.

4.6 Outliers and influential points in simple linear regression

Unusual values in simple linear regression can occur in the response (or dependent) variable Y, or in the regressor (or independent, or predictor) variable X, or both, with differing consequences (Fig. 4.6). An observation with an unusual regressor value is potentially very valuable in determining the slope of the fitted least squares line, and so is given more weight in determining the line than the other data values. Such a point is termed a leverage point because it exerts a strong pull on the regression line as if it were pulling a lever. Leverage points are not necessarily bad; they can really 'pin down' the regression line and so result in dramatically improved standard errors for the slope estimate and other estimates calculated in the regression. However,

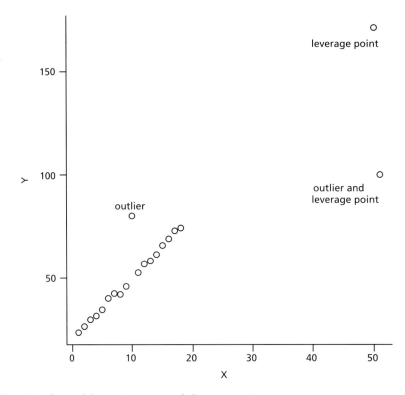

Fig. 4.6 Influential data points in a simple linear regression.

since a leverage point exerts an undue influence on the regression, it is best to be aware of its existence and to double-check its values.

Leverage is determined entirely by an observation's regressor (X) value. In the regression setting, an outlier is a different kind of unusual value, an observation with a response (Y) value that does not fit the X–Y pattern found in the rest of the data. Note that this can happen even if neither the X value nor the Y value is unusual in its own right. Outliers in regression can have undue influence on the fitted regression, though they do not always do this. They will certainly reduce the regression R^2 value and inflate the error mean square, and this results in inflated standard errors for all estimated quantities and loss of power for all hypothesis tests. Of course, the same point in a regression data set can be both a leverage point and an outlier (Fig. 4.6).

4.6.1 Diagnostic measures for leverage points and outliers

Diagnostic measures have been defined to measure and detect leverage points and outliers (Belsley et al. 1980). For example, the leverage of the ith data point is defined to be

$$h_i = (1/n) + (X_i - \bar{X})^2 / (n-1)S^2_X$$

where $i = 1, 2, \ldots, n$, and $\bar{X}$ and S^2_X are the mean and variance of the regressor, respectively. (Leverage values can be obtained from the PROC REG procedure of SAS (SAS 1990) by requesting their inclusion in an output data set.) In a simple linear regression, the average of these h_i values is $2/n$, and the ith data point is a leverage point (under some conventions) if $h_i > 4/n$. Under one set of assumptions, about one in 20 points will qualify 'accidentally' as a leverage point using this criterion, so some analysts prefer the more stringent cutoff value $6/n$, for which the 'clean-data' incidence is on the order of 1 in 100.

Outliers in regression can be detected by means of 'studentized' residuals. Several varieties have been defined, but the externally studentized residual is recommended:

$$r_i = e_i / \sqrt{MSE_{(-i)}(1 - h_i)}$$

where e_i is the ith ordinary residual (actual Y_i − predicted Y_i) and $MSE_{(-i)}$ is the error mean square for the regression excluding the ith pair. If the assumptions of the simple linear regression model hold, studentized residuals can be used to test for contamination, since each r_i follows a Student's t-distribution with $(n - 3)$ degrees of freedom under the hypothesis of no contamination. Hence, a two-sided test would assert contamination if $|r_i| > t_{\alpha/2, n-3}$, the upper-$\alpha/2$ critical point from the t distribution with $n - 3$ degrees of freedom. In this case, α is a per-observation error rate, and should again usually be set lower than 0.05. For example, in a perfectly 'clean' data set containing 100 points, we expect five studentized residuals to exceed the $\alpha = 0.05$ critical value, and one to exceed the $\alpha = 0.01$ value purely by accident. Some authors have suggested using the Bonferroni inequality, which at the 0.05 level means using $t_{\alpha/2n, n-3}$ as a cutoff for each studentized residual; this will often be an extremely conservative approach. (Studentized residuals can also be obtained from the PROC REG procedure of SAS (SAS 1990) by requesting inclusion of RSTUDENT values in the output data set.)

4.6.2 An example

To increase awareness of the role of influential data in the simple linear regression setting, consider the data of Allison and Cicchetti (1976) on 63 species of terrestrial mammals shown in Fig. 4.7. In any study comparing brain weights of animal species, some correction should be made for body weight. One approach to doing this would be to regress brain weight (Y) on body weight (X) and use the residuals as a measure of adjusted brain weight. In the regression shown in Fig. 4.7, both the Asian elephant ($h = 0.1279$) and African elephant ($h = 0.8612$) are leverage points. For the data shown in Fig. 4.7, the $\alpha = 0.01$ critical point for the outlier test is $t_{0.005, 59} = 2.657$, and both of the

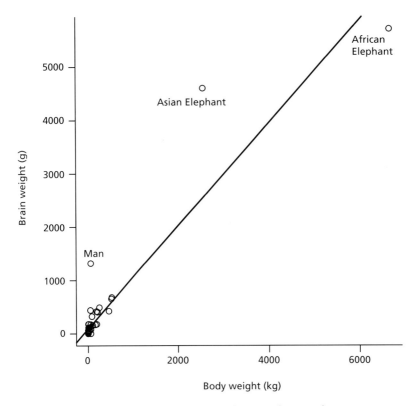

Fig. 4.7 Brain weight vs. body weight, 63 species of terrestrial mammals.

elephants (Asian r = 12.30; African r = −11.85) and Man (r = 3.95) 'flunk' the outlier test.

Of course, these outlier tests are only valid if the assumptions of the regression hold. These assumptions are:
1 The values of the regressor X are known constants (measured with negligible error).
2 At any fixed X, the long-run mean of many Y-values, say μ(X), is a linear function of X.
3 The regression errors (the deviations of repeated Y-values at a given X from their long-run mean μ(X)) are normally distributed, with constant variance, and are independent.

For the data of Fig. 4.7, several of these assumptions are either questionable or difficult to assess. The assumption of linearity cannot be verified for body weights beyond 1000 kg, since there are so few points at these values. The assumption of constant error variance probably doesn't hold with so many points 'packed' into the lower left corner of the plot.

Both of these variables vary over several orders of magnitude, and no analysis of the raw data will distinguish among the lower orders of magnitude. As

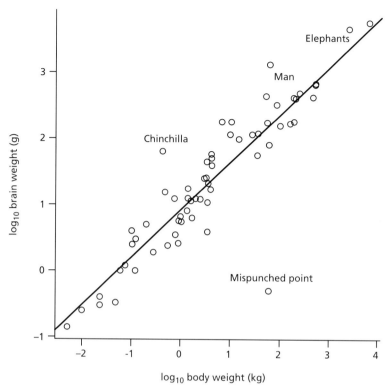

Fig. 4.8 Logged brain weight vs. logged body weight.

long as elephants are included in the data, the baboons, lemurs, and field mice will all seem equal in size (will all seem to be 0) unless the analysis is done on an order-of-magnitude scale: the log scale. Figure 4.8 shows a plot of these data in the log scale, $Y^* = \log_{10}(\text{brain wt})$ versus $X^* = \log_{10}(\text{body weight})$. When checked carefully, the formal assumptions of the regression appear to hold, with the possible exception of some points whose Y^* values do not fit the pattern (i.e. possible outliers). There are no leverage points, but the point at lower right in Fig. 4.8, labelled 'mispunched point', is a severe outlier. Its studentized residual value is $r^* = -7.56$. The point was placed in this data set for the purposes of demonstration. Although an obvious outlier in Fig. 4.8, it is also present (but undetectable) in the untransformed data of Fig. 4.7. It is also undetectable using univariate outlier tests such as Grubbs' test, since both its X-value and Y-value are individually well within the range of other values found in the data. This point is the promised example of a model-dependent outlier.

Upon removal of the mispunch and reanalysis, two other points in this data set emerge as possible outliers. Man ($r^* = 2.670$) barely signals using $\alpha = 0.01$, but the chinchilla's brain weight ($r^* = 3.785$) is highly unusual given its body weight.

4.7 Influential data in multiple linear regression

In simple linear regression, formal diagnostic measures for leverage points and outliers are almost unnecessary, since these points are generally noticeable by eye in a plot of the response variable versus the regressor. In multiple linear regression this is no longer true. In this case, the diagnostic checks using leverage values and studentized residuals can help a data analyst find influential observations that are well hidden in scatter plots and other simple inspection tools.

Suppose the response variable Y is to be regressed on the p − 1 regressor variables $X_1, X_2, \ldots, X_{p-1}$, in a model containing an intercept, and therefore requiring estimation of p unknown regression coefficients, $\beta_0, \beta_1, \ldots, \beta_{p-1}$. Here, the three formal statistical model assumptions are the same as given in section 4.6.2, except for the following change to the second assumption: 2*. At any fixed $X_1, X_2, \ldots, X_{p-1}$, the long-run mean of many Y values, say $\mu(X)$, is a linear function of these values, $\mu(X) = \beta_0 + \beta_1 X_1 + \ldots + \beta_{p-1} X_{p-1}$.

Even with only two regressors, the model $Y = \beta_0 + \beta_1 X_1 + \beta_2 X_2 + \varepsilon$ relates the values of three variables. A scatter plot of Y versus X_1 can be considered a 'side-view' of points on a three-dimensional surface. Because this plot cannot show depth, it quite possibly gives misleading impressions of the true relationships in the data. Also, it is quite possible for outliers and leverage points to hide in each two-dimensional scatter plot in much the same way the mispunched data point of Fig. 4.8 hides in separate histograms of X and Y.

4.7.1 Diagnostic measures for leverage points and outliers

When more than one regressor is to be used, the concept of leverage alters slightly; an observation is now influential in this sense if it has an unusual *combination* of predictor-variable values. That is, no single predictor variable need have an outlying value; it is the combination of variables that's important. Formally, if X represents the n × p matrix whose first column is all ones, second column the values of X_1, third column the values of X_2, etc., the leverages h_i, i = 1, 2, ..., n of the observations are the diagonal elements of the matrix $H = X(X'X)^{-1}X'$. An observation is considered a leverage point if its h_i value exceeds 2p/n, or more conservatively 3p/n. Surprisingly, the definition of studentized residuals, so useful for detecting outliers in simple linear regression, does not change (formula supplied in section 4.6.1), though the definitions of its component parts change, in obvious ways. As before, an outlier is an observation whose response value does not fit the X–Y relationship pattern found in the rest of the data. In a multiple regression with p coefficients estimated, the studentized residuals should be compared to a Student's t critical points with n − p − 1 degrees of freedom for a formal outlier test.

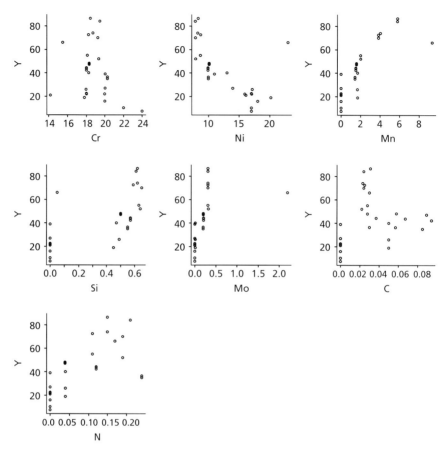

Fig. 4.9 Preliminary scatter plots of Neel temperature of stainless steel coupons vs. elemental percentages.

4.7.2 An example

Though not an ecological data example, the study of Jones *et al.* (1987) is a true scientific story that dramatically illustrates several of the above points. In this study of low-temperature magnetic properties of stainless steel, the subjects were n = 29 steel coupons with varying elemental percentages (other than the base element iron) of chromium (Cr), nickel (Ni), manganese (Mn), silicon (Si), molybdenum (Mo), carbon (C), and nitrogen (N). The response variable (Y) for each coupon was its Neel temperature, the temperature in °K below which the magnetic properties of the steel no longer obey well-known laws of physics. One goal of the analysis was to build a regression model that could predict Neel temperature for future steel coupons in terms of the percentages of these seven elements. Figure 4.9 shows pre-plots of the Neel temperature (Y) versus each of the seven predictors. These plots belie the strength

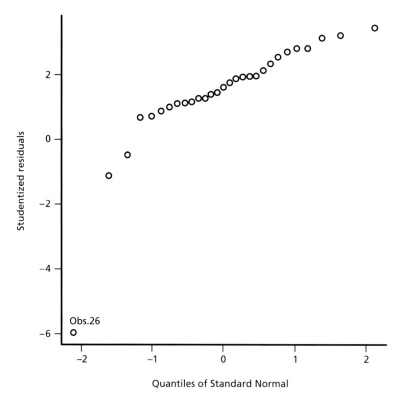

Fig. 4.10 Normal probability plot of studentized residuals, Neel temperature full data set regression.

of the true relationships in the data; the reader will be surprised to learn that a multiple regression fit to the full data set has an R^2 of 0.9903! All predictors except Si, Mo and C are highly statistically significant based on partial t-tests. The lesson to be learned from this data set is that a high R^2 does not necessarily mean the analysis is complete.

The pre-plots belie the strength and nature of the actual relationships, and they are also misleading in terms of identifying influential points. For example, based on these plots, one is led to believe that the rightmost points on the Ni, Mn, and Mo plots are outliers, but they are not. What's more, they all correspond to the same coupon. Calculation of leverage values shows that observations 25 and 26, with h-values 0.993 and 0.576, respectively, are leverage points using the mild $2p/n = 16/29 = 0.55$ cutoff. The more stringent $3p/n$ rule ($24/29 = 0.83$) flags only observation 25. This observation is the one to the right in the Ni, Mn, and Mo plots, but its studentized residual is only -0.3969. This means that observation 25, though it has a highly unusual combination of elemental percentages, still gives a Y value that fits the pattern found in the rest of the data: it is a leverage point, but not an outlier.

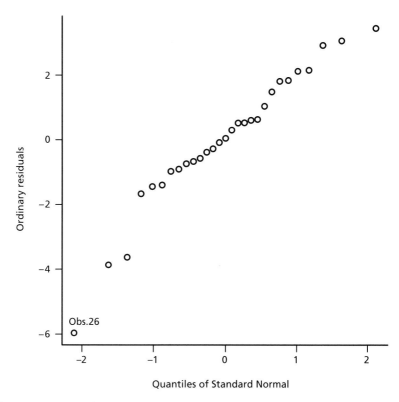

Fig. 4.11 Normal probability plot of ordinary residuals, Neel temperature full data set regression.

There is an outlier in this regression, it is observation 26. There is little or no indication in any of the preplots of Fig. 4.9 that this observation is an outlier. It is well within the bulk of the points in every plot except the Cr pre-plot; in that one, it is the fairly innocuous point at lower left. A plot of the studentized residuals (Fig. 4.10) for the full regression exhibits an approximate straight line for the bulk of the data, with one point extremely out of place at lower left. This point's value, $r_{26} = -6.458$, is well beyond the $\alpha = 0.01$ critical point $t_{0.005,20} = 2.77$ in absolute value and would be extremely improbable if observation 26 were uncontaminated. A similar analysis using ordinary residuals instead of studentized residuals shows observation 26 as being slightly unusual but not overwhelmingly so (Fig. 4.11).

When notified of the possible contamination in observation 26, the physicists went to their lab records and found an error in their procedures that justified removal of this point from the regression. Once removed, the R^2 jumped from 0.9903 to 0.9968, corresponding to a nearly threefold reduction in the error sum of squares and error mean square. The normal plot of studentized residuals for this reduced-data regression is 'textbook clean' (Fig. 4.12).

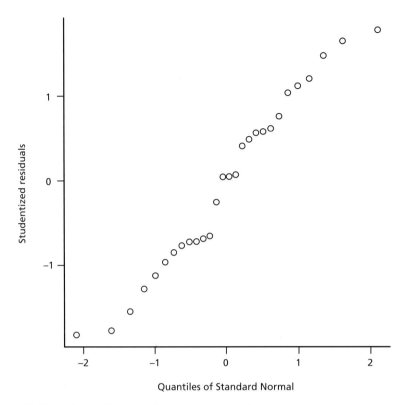

Fig. 4.12 Normal probability plot of studentized residuals, Neel temperature regression excluding obs. 26.

The reduction in mean square for error led to all regressors becoming statistically significant at $p < 0.001$; this was an important finding for the physicists, since prior studies had suggested carbon was unimportant in predicting Neel temperature. The final regression model was capable of predicting Neel temperatures with an accuracy of $\pm 3°K$.

As a postscript, in the regression excluding observation 26, observation 25 still qualifies as a leverage point. It is an example of a 'good' influential point, in that, when it is removed from the data, the regression coefficient for Mo switches signs and becomes statistically insignificant. Apparently, because of its very large percentage of Mo, observation 25 is crucial for estimation of the regression coefficient for this elemental percentage.

4.8 Conclusions

Detailed discussion has been offered concerning the prevention and detection of contamination in samples and in regression. Much more could be said. For example, Grubbs' test can be adapted to the setting of repeated small samples, as would often be the case in water quality studies, by using a pooled variance estimator over several samples. There are also different versions of the test if one suspects there is more than one outlier in the sample. Also not discussed is the problem of instrument miscalibration, which may result in a number (possibly large) of outliers; these are actually variable values shifted by an additive and/or multiplicative constant. Entire volumes have been written on calibration, and new research results continue to emerge.

All of the techniques presented in this chapter work well for small- to moderate-sized data sets. Very large data sets are becoming increasingly commonplace, and will require new methods for quality assurance, or creative adaptations of existing methods. One obvious example is that the choice of α–value for outlier detection using studentized residuals should be made even smaller; for example, using $\alpha = 0.01$, we expect to naturally find many 'ouliers' in an uncontaminated data set comprised of thousands (or millions) of observations.

In two of the real data examples in this chapter, extreme values seemed to exist in the raw values but not in the log-transformed values. In this author's experience, there are more log-normally distributed data in nature than normally distributed data. With astonishing frequency, measurement processes that seem to spawn frequent outliers in the raw data scale show clean, regular, unremarkable variation in the log scale, or more generally after some power transformation (Box & Cox 1964). If the data cross several orders of magnitude (i.e. if the ratio of maximum Y to minimum Y is greater than 10), it is highly recommended that analyses be performed on (at least) both the raw data and the log-transformed data. Patterns will often be more linear, and variation more homoscedastic and Gaussian, in the log scale. Moreover, meaning is not really lost, and in some cases the conclusions are more meaningful using a log-transformed response variable (because treatment and predictor variable effects are expressed in relative terms, e.g. percentage increase, when the response variable is log-transformed).

Finally, no detailed discussion of modern robust statistical methods such as Iteratively Reweighted Least Squares (IRLS) algorithms has been offered here (see Little 1990). For example, Fig. 4.13 shows the fit of the ordinary least squares line (dotted) and a robust regression line (solid) using the Splus function rreg. The least squares line is affected quite adversely by the combination leverage point-outlier at lower right. The robust regression fit begins with an ordinary least squares line, and then refits a weighted least squares line with weights inversely related to the absolute residuals of the first fit. After several iterations of this reweighting and refitting scheme, serious outliers are often downweighted to the point of being removed from the data. Robust regression procedures could, in some cases, be considered automatic outlier-detection algorithms. However, the danger of mindless dependency on automatic

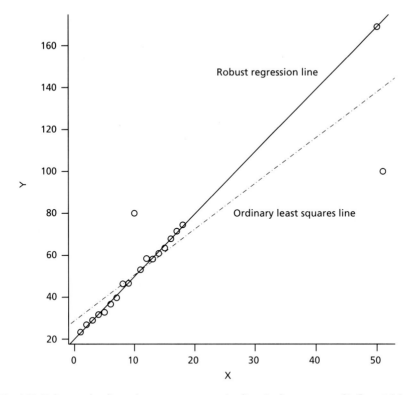

Fig. 4.13 Robust and ordinary least squares regression lines in the presence of influential data.

detection-elimination algorithms is worrisome to this author despite the sophisticated mathematical and numerical techniques used to fit robust regression models. These algorithms are potentially very useful in the right hands and are a topic of active research.

4.9 References

Allison, T. & Cicchetti, D.V. (1976) Sleep in mammals: ecological and constitutional correlates. *Science* **194**, 732–734.
ASTM E 178–94. (1994) *Standard Practice for Dealing with Outlying Observations*. American Society for Testing and Materials, Philadelphia, PA.
Belsley, D.A., Kuh, E. & Welsch, R.E. (1980) *Regression Diagnostics: Identifying Influential Data and Sources of Collinearity*. John Wiley, New York.
Blom, G. (1958) *Statistical Estimates and Transformed Beta Variables*. John Wiley, New York.
Box, G.E.P. & Cox, D.R. (1964) An analysis of transformations. *Journal of the Royal Statistical Society* **B26**, 211–243, discussion 244–252.
Chambers, J.M., Cleveland, W.S., Kleiner, B. & Tukey, P.A. (1983) *Graphical Methods for Data Analysis*. Duxbury Press, Boston, MA.
Daniel, C. (1959) Use of half-normal plots in interpreting factorial two-level experiments. *Technometrics* **1**, 311–341.

Deming, W.E. (1986) *Out of the Crisis.* Massachusetts Institute of Technology, Center for Advanced Engineering Study, Cambridge, MA.

Flournoy, N. & Hearne, L.B. (1990) Quality control for a shared multidisciplinary database. In: *Data Quality Control: Theory and Pragmatics.* (eds G.E. Liepins & V.R.R. Uppuluri), pp. 43–56. Marcel Dekker, New York.

Grubbs, F.E. & Beck, G. (1972) Extension of sample sizes and percentage points for significance tests of outlying observations. *Technometrics* **14**, 847–854.

Jones, E.R., Datta, T., Almasan, C., Edwards, D. & Ledbetter, H.M. (1987) Low-temperature magnetic properties of FCC Fe-Cr-Ni alloys: effects of manganese and interstitial carbon and nitrogen. *Materials Science and Engineering* **91**, 181–188.

LePage, N.J. (1990) Data quality control at United States Fidelity and Guaranty Company. In: *Data Quality Control: Theory and Pragmatics.* (eds G.E. Liepins & V.R.R. Uppuluri), pp. 25–41. Marcel Dekker, New York.

Little, R.J. (1990) Editing and imputation of multivariate data: issues and new approaches. In: *Data Quality Control: Theory and Pragmatics.* (eds G.E. Liepins & V.R.R. Uppuluri), pp. 145–166. Marcel Dekker, New York.

SAS Institute Inc. (1990) *SAS/STAT User's Guide.* SAS Institute, Inc., Cary, North Carolina.

Simpson, J., Olsen, A. & Eden, J.C. (1975) A Bayesian analysis of a multiplicative treatment effect in weather modification. *Technometrics* **17**, 161–166.

Sokal, R.R. & Rohlf, F.J. (1995) *Biometry.* W.H. Freeman and Co., New York.

Stolarski, R.S., Krueger, A.J., Schoeberl, M.R., McPeters, R.D., Newman, P.A. & Alpert, J.C. (1986) Nimbus 7 satellite measurements of the springtime Antarctic ozone decrease. *Nature* **322**, 808–811.

CHAPTER 5

Metadata

WILLIAM K. MICHENER

5.1 Introduction

Most data sets used by ecologists and other scientists comprise rows and columns of numbers and characters that represent measurements collected in the field and laboratory. The structure and content of the rows and columns reflect how the scientist(s) thinks about data, the experimental design, sampling methods, temporal context and other information that is seldom written down as the data are prepared for analysis and interpretation. Few data sets, whether written in a notebook or entered into a spreadsheet or database program, are perfect and self-explanatory. Ecologists must, therefore, rely upon memory, field notes or other forms of documentation which allow them to understand and use a specific data set, as well as correctly interpret results from subsequent processing and analysis. Metadata are the information necessary to understand and effectively use data, including documentation of the data set contents, context, quality, structure and accessibility.

Metadata are receiving increased attention from the scientific community. Ecologists, scientific societies and state and federal agencies are recognizing the importance of high quality, well-documented and securely archived data for addressing long-term and broad-scale environmental questions (e.g. Gross *et al.* 1995; NRC 1995). Furthermore, ecological and environmental data, such as those collected by individuals and teams of scientists at field stations, marine laboratories, parks, natural areas and preserves, represent significant institutional, regional, national and international resources essential for understanding and monitoring the health of the dynamically changing environment. Comprehensive metadata comprise the key that can 'unlock' these data resources, thereby supporting broad and long-term use and interpretation.

The objectives of this chapter are to:
1   examine some of the benefits and costs associated with developing and implementing metadata;
2   review metadata content 'standards' that may be relevant to ecologists;
3   identify metadata software and resources;
4   outline issues that warrant attention during the implementation process; and
5   identify remaining challenges and opportunities related to metadata implementation.

METADATA 93

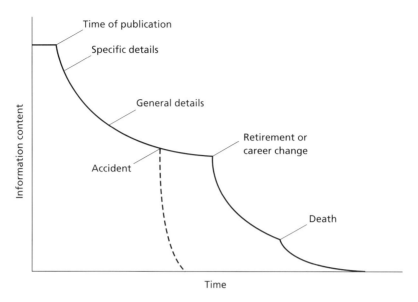

Fig. 5.1 Example of the normal degradation in information content associated with data and metadata over time ('information entropy'). Accidents or changes in storage technology (dashed line) may eliminate access to remaining raw data and metadata at any time. (After Michener *et al.* 1997, by permission of the Ecological Society of America.)

5.2 Benefits and costs associated with metadata

The rows and columns of numeric and textual observations contained within a data set are frequently referred to as raw data. These raw data are considered useful when they can be processed within the scientific framework of the study that served as the genesis for the data. Relating raw data to a study's underlying theoretical or conceptual model(s), in such a way that the data can be used and interpreted, requires understanding the types of variables measured, the context under which they were measured, measurement units, data quality and other pertinent facts not represented in the raw data (Michener *et al.* 1997). This material is provided in metadata. Information then results from interpretation of the combination of raw data and metadata within a study's conceptual framework.

Information content can be lost through the degradation of the raw data or the metadata. Such loss is unavoidable and has been referred to as information entropy (Michener *et al.* 1997). Many processes can lead to information entropy (Fig. 5.1) including the continuous, gradual deterioration of storage media. More often discrete events, including career changes, retirement, or death of the scientist(s) and technician(s) involved in data collection, obsolescence of storage technology or analytical equipment, and loss of storage media through catastrophic events, may result in partial or total loss of data at any point in time (Bowser 1986; Michener *et al.* 1997). Although metadata loss and degradation can occur throughout the period of data collection and analysis, the rate of loss frequently increases after project results have been published or the study has been terminated. Specific details are usually lost first due to abandonment or loss of data forms and notes and/or to the investigator's memory attrition. Over longer periods, storage media degeneration and further memory losses can reduce the amount of available, detailed information not included in publications.

5.2.1 Benefits

Benefits and costs accrue during the development, implementation and maintenance of metadata. At least three major benefits may be derived as a consequence of investing adequate time and energy into metadata development (Michener *et al.* 1997). First, data entropy is delayed and, correspondingly, data set longevity is increased. As a consequence of data complexity, time and funding constraints, and information entropy, the life span of a typical ecological data set may be very short, possibly lasting only from data set conception to publication. Even when data are properly archived and maintained, history and personal experiences provide numerous examples of data that became useless because relevant metadata were missing or unavailable (NRC 1995; Michener *et al.* 1997). Development and maintenance of comprehensive metadata can counteract this natural tendency for data to degrade in information content through time.

Second, data reuse by the originator and data sharing with others are facilitated. Human memory is limited. With the possible rare exception of extremely simple data sets that are analysed immediately after collection, even data originators require some metadata to support subsequent processing and analysis. Moreover, scientists require highly detailed instructions or documentation in order to analyse and interpret historic or long-term data sets accurately, as well as data from unfamiliar research or complicated experimental designs.

Third, well-documented data may be used to expand the scale of ecological inquiry and support valid comparisons in space and time. For example, short-term studies may be integrated into or evolve into long-term studies. Metadata are essential for maintaining a historical record of long-term data sets that have resulted from such integration efforts, as well as documenting data anomalies and changes in personnel, methods and instrumentation in ongoing long-term, projects. Furthermore, metadata are critical for combining physical, chemical and biological data sets containing different parameters but sharing common spatial or temporal domains. Thus, comprehensive metadata can enable data sets that were designed for a single purpose to be used for multiple purposes, often repeatedly over long periods. In many cases, the scientific value of being able to reuse data and to utilize data for multiple, often unforeseen, objectives may far exceed the perceived value associated with publications resulting from the original study. For example, Taylor (1989) and Leigh and Johnston (1994) include numerous case studies that illustrate how long-term data sets like those initiated during the mid 1800s at Rothamsted Experimental Station (Harpenden, UK) continue to be used to expand our ecological understanding.

Other benefits of metadata may include satisfaction of funding agency and organizational mandates. For example, funding agencies and other organizations (e.g. governmental agencies) often require or recommend comprehensive metadata, especially for studies that may be brought into litigation, as well as long-term, high-cost and high visibility studies.

5.2.2 Costs

High costs, primarily in terms of personnel time, can be associated with developing and maintaining metadata. For example, the volume of metadata and the level of effort expended in developing metadata for relatively simple, short-term experiments may exceed the physical size of the raw data file (possibly by several orders of magnitude) and the effort expended in data collection (Porter *et al.* 1997). This is not unusual in disciplines such as chemistry and physics where understanding the experimental conditions is critical for repeatability. In such cases, metadata may be scrutinized as much, if not more than the data. Real costs are associated with editing data and metadata and making them available in paper or electronic formats to the scientific community. Moreover, costs associated with developing and disseminating metadata are rarely included in project budgets.

Long-term stewardship and maintenance of data and metadata also represent real cost burdens that often are not factored into project and institutional budgets. Furthermore, myriad potential secondary uses for a particular data set, each possibly requiring comprehensive metadata for a particular aspect of the data set, may not be anticipated. Critical details can be overlooked easily in even the most comprehensive metadata. Consequently, the provision for archiving data and metadata does not necessarily guarantee the long-term utility of a data set. Moreover, after a study has been terminated, informing the user community of changes to the data set and of newly discovered anomalies is difficult. Direct consultation with the data originator(s) may, whenever possible, be necessary to resolve questions and issues that arise during secondary data use.

5.3 Metadata content 'standards' relevant for ecology

All ecological data are collected at one or more locations and, consequently, have a spatial or geographic component. The spatial component of the data may range from being central to being relatively unimportant to the success of a project. Geospatial data, for example, are explicitly associated with multiple geographical locations. In such cases, both the environmental attributes associated with each sampling point and the specific location of the points are of scientific interest. Examples include remotely sensed imagery, geographic information system (GIS) data layers, data derived from broad-scale sampling efforts (e.g. meteorological networks), as well as fine-scale sampling of spatially explicit patterns and processes. Most metadata standardization efforts have focused, thus far, on data having a strong geospatial component.

In contrast, non-geospatial data might include data from laboratory and microcosm to mesocosm experiments, as well as other ecological data collected at a finite number of points. For these cases, which probably comprise most ecological studies, precise geographic coordinates of the sampling sites may be relatively unimportant or not recorded. Less attention has been paid to developing approaches for comprehensively documenting these types of studies. Although the distinction between geospatial and non-geospatial data is, admittedly, artificial, the two categories are treated independently in the following discussion. This is done primarily to emphasize the breadth of information and level of detail that are essential for understanding and replicating ecological studies.

5.3.1 Geospatial metadata

Significant effort has been expended in developing geospatial metadata standards during the past decade. Many of the spatial data standards developed in the early 1990s in the USA incorporated a metadata component (Digital Geographic Information Working Group 1991; Defense Mapping Agency 1992; National Institute of Standards and Technology 1992). More recently, the Federal Geographic Data Committee (1994, 1998) completed the Content Standards for Digital Geospatial Metadata. These standards were developed as part of both the ongoing evolution of the National Biological Information Infrastructure in the United States and the standardization of geographical data collected by the Federal Government. The Content Standards contain more than 200 metadata fields that are categorized into seven classes of metadata descriptors: identification, data quality, spatial data organization, spatial reference, entity and attribute, distribution, and metadata. Efforts are underway to add extensions to the Content Standards, creating metadata supersets relevant to biological, vegetation classification, cultural, demographic and other types of data (see Stitt & Nyquist 1996 for examples).

Similar standardization efforts are taking place throughout the world. For instance, metadata guidelines have been produced for use in the United Kingdom by the National Geospatial Data Framework (NGDF) Management Board (NGDF 1998). For more information about emerging metadata standards in Europe, contact MEGRIN (http://www.megrin.org/), an organization representing and owned by National Mapping Agencies from throughout Europe. MEGRIN also maintains a Geographic Data Description Directory on the World Wide Web that provides information about digital map data available in European countries.

5.3.2 Non-geospatial ecological metadata

The role of metadata in facilitating ecological research has been recognized since the 1980s (Kellogg Biological Station 1982; Stafford *et al.* 1986; Michener *et al.* 1987; Kirchner *et al.* 1995). However, metadata standards for non-geospatial ecological data do not exist currently in any accepted format beyond individual studies, projects, or organizations. Ecological studies often require large amounts of a disparate data related to the chemical and physical attributes of the environment, as well as the individual organisms, populations, communities and ecosystems composing the biotic portion of the environment. It is unlikely that a single metadata standard, no matter how comprehensive, could encompass all types of ecological data.

Consequently, a generic set of non-geospatial metadata descriptors was recently introduced for the ecological sciences (Michener *et al.* 1997). The list of metadata descriptors was proposed as a template that could serve as the basis for more refined subdiscipline- or project-specific metadata guidelines. Five classes of metadata descriptors were delineated.

1 data set descriptors—basic attributes of the data set (e.g. data set title, associated scientists, abstract and keywords).

2 research origin descriptors—all relevant metadata that describe the research leading to the genesis of a particular data set (i.e. hypotheses, site characteristics, experimental design and research methods).

3 data set status and accessibility descriptors—the status of the data set and associated metadata, as well as information related to data set accessibility.

4 data structural descriptors—all attributes related to the physical structure of the data file.

5 supplemental descriptors—all other related information that may be necessary for facilitating secondary usage, publishing the data set or supporting an audit of the data set (Table 5.1).

Table 5.1 Ecological metadata descriptors. (After Michener *et al.* 1997, by permission of the Ecological Society of America.)

Class I: Data set descriptors
A Data set identity — Title or theme of data set
B Data set identification code — Accession number or code specified by the data originator or data management personnel to uniquely identify a data set
C Data set description
  1 originator(s) — Name(s) and address(es) of principal investigator(s) associated with data set
  2 abstract — Summary of research objectives, data set contents (including temporal and spatial context), and potential uses of the data set
D Keywords — Theme and contents, location (including ecosystem type)

Class II: Research origin descriptors
A Overall project description — *Note*: this section may be essential if data set represents a component of a larger or more comprehensive database; otherwise, relevant items may be incorporated into II B
  1 identity — Project title or theme
  2 originator(s) — Name(s) and address(es) of principal investigator(s) associated with project
  3 study period — Start date and end date (or expected duration)
  4 objectives — Scope and purpose of research program
  5 abstract — Summary of the overall research project
  6 source(s) of funding — Name(s) and address(es) of funding sources, grant and contract numbers, and funding period

B Sub-project description
  1 site description
    (a) location — Latitude and longitude, political geography, permanent landmarks and reference points
    (b) physiographic region — Ecoregion or physiographic province
    (c) landform component — Ridge, backslope, flood plain, stream terrace
    (d) watershed(s) — Size, boundaries, receiving streams and rivers
    (e) topographic attributes — Elevation, slope, aspect, microtopography
    (f) geology, lithology and soils — Soil parent material type, lithology of predominant soil parent material, geomorphic history and approximate age of geomorphic features, soil taxonomic units (e.g., order and series; see Robertson *et al.* 1999)
    (g) vegetation communities — Descriptive (e.g., short-grass prairie, blackwater stream, etc.) and detailed (e.g., associations, species lists) when appropriate
    (h) history of land use and disturbances — History of site management practices (e.g., burning, cutting, grazing) and natural and anthropogenic disturbances (e.g., fire, floods, pest outbreaks)
    (i) climate — Summary of site climatic characteristics
  2 experimental or sampling design
    (a) design characteristics — Description of statistical/sampling design, details pertaining to subsampling and replication
    (b) permanent plots — Dimension, location, general vegetation characteristics (if applicable)
    (c) data collection duration and frequency — Information necessary to understand temporal sampling regime

Contd

Table 5.1 (*contd*)

3	research methods	
	(a) field and laboratory	Description or reference to standard field/laboratory methods
	(b) instrumentation	Description, manufacturer, and model/serial numbers
	(c) taxonomy and systematics	References for taxonomic keys, identification and location of voucher specimens, etc.
	(d) permit history	References to pertinent scientific and collecting permits
	(e) legal/organization requirements	Relevant laws, decision criteria, compliance standards, etc.
4	project personnel	Principal and associated investigator(s), technicians, supervisors, students

Class III: Data set status and accessibility

A	Status	
	1 latest update	Date of last data set modification
	2 latest archive date	Date of last data set archival
	3 metadata status	Date of last metadata update and current status
	4 data verification	Status of data quality assurance checking
B	Accessibility	
	1 storage location and medium	Pointers to where data reside (including redundant archival sites)
	2 contact person(s)	Name, address, phone, fax, electronic mail address, and WWW page
	3 copyright restrictions	Whether copyright restrictions prohibit use of all or portions of the data set
	4 proprietary restrictions	Any other restrictions which may prevent use of all or portions of the data set
	(a) release date	Date when proprietary restrictions expire
	(b) citation	How data may be appropriately cited
	(c) disclaimer(s)	Any disclaimers which should be acknowledged by secondary users
	5 costs	Costs associated with acquiring data (which may vary by size of data request, desired medium, etc.)

Class IV: Data structural descriptors

A	Data set file	
	1 identity	Unique file names or codes
	2 size	Number of records, record length, number of bytes
	3 format and storage mode	File type (e.g., ASCII, binary), compression schemes employed (if any), etc.
	4 header information	Description of any header data or information attached to file {Note: header information may include elements related to 'variable information' (IV. B.) and, if so, could be linked to the appropriate section(s)}
	5 alphanumeric attributes	Mixed, upper, or lower case
	6 special characters/fields	Methods used to denote comments, 'flag' modified or questionable data, etc.
	7 authentication procedures	Digital signature, checksum, actual subset(s) of data, and other techniques for assuring accurate data transmission to secondary users
B	Variable information	
	1 variable identity	Unique variable name or code
	2 variable definition	Precise definition of variables in data set

Contd p. 100

Table 5.1 (*contd*)

3 units of measurement	SI units of measurement associated with each variable
4 data type	
(a) storage type	Integer, floating point, character, string, etc.
(b) list and definition of variable codes	Description of any codes associated with variables
(c) range for numeric values	Minimum, maximum
(d) missing value codes	Description of how missing values are represented in data set
(e) precision	Number of significant digits
5 data format	
(a) fixed, variable length	
(b) columns	Start column, end column
(c) optional number of decimal places	
C Data anomalies	Description of missing data, anomalous data, calibration errors, etc.

Class V: Supplemental descriptors

A Data acquisition	
1 data forms or acquisition methods	Description or examples of data forms, automated data loggers, digitizing procedures, etc.
2 location of completed data forms	Physical location (address, building, room number)
3 data entry verification procedures	Procedures employed to verify that digital data set is error-free
B Quality assurance/quality control procedures	Identification and treatment of outliers, description of quality assessments, calibration of reference standards, equipment performance results, etc.
C Supplemental materials	References and locations of maps, photographs, videos, GIS data layers, physical specimens, field notebooks, comments, etc.
D Computer programs and data processing algorithms	Description or listing of any algorithms or software used in deriving, processing or transforming data
E Archival	
1 archival procedures	Description of how data are archived for long-term storage and access
2 redundant archival sites	Locations and procedures followed to provide redundant copies as a security measure
F Publications and results	Electronic reprints, listing of publications resulting from or related to the study, graphical/statistical data representations, primary WWW site(s) for data and project
G History of data set usage	
1 data request history	Log of who requested data, for what purpose, and how it was actually used
2 data set update history	Description of any updates performed on data set
3 review history	Last entry, last researcher review, etc.
4 questions and comments from secondary users	Questionable or unusual data discovered by secondary users, limitations or problems encountered in specific applications of the data, unresolved questions or comments

The metadata descriptors were formulated to answer five basic questions that might arise when an ecologist attempts to identify and use a specific data set:
1   What relevant data exist?
2   Why were those data collected and are they suitable for a particular use?
3   How can these data be obtained?
4   How are the data organized and structured?
5   What additional information is available that would facilitate data use and interpretation?

It can be especially difficult to identify and document all supplemental information that may be required for specific data uses. For this reason, it may be beneficial to design metadata that can also serve as a vehicle for user feedback and data anomaly reporting. A 'data set usage history' (Table 5.1, Class V., G), for example, may add value to data sets and facilitate their long-term use. This activity is synonymous to attaching 'Post-it Notes' to the data to alert subsequent users to idiosyncrasies within the metadata or to anomalies within the data (Michener *et al.* 1997).

5.3.3 Related metadata activities and standards

Several generic metadata 'standards' have been developed by information scientists to facilitate cataloguing and discovery of electronic resources. Some examples are the Dublin Core (Desai 1997; Weibel 1997; Thiele 1998); NASA's Directory Interchange Format (DIF; NASA Goddard Space Flight Center 1993); and the Government Information Locator Service format (GILS; see Barton 1996). Metadata descriptors included in each of the standard formats are listed in Table 5.2. These formats represent the least common denominators for resource descriptions and vary according to the objectives of the institution or workshop (e.g. Dublin Core Workshop). Additional metadata formats (i.e. IAFA/Whois++, MARC, Text Encoding Initiative, Uniform Resource Characteristics) are reviewed by Heery (1996). These generic formats are extremely useful for facilitating cataloguing and discovery of resources but do not obviate the need for more comprehensive metadata descriptions (e.g. FGDC 1994, 1998; Michener *et al.* 1997).

Several additional comprehensive, discipline-specific metadata formats have been proposed or are under development. For example, the generic list of ecological metadata descriptors developed by Michener and colleagues (1997) was modified as part of an effort to develop standard soil methods for long-term ecological research (Boone *et al.* 1999). Metadata content guidelines for climate monitoring, emphasizing those descriptors necessary for determining data fitness-for-use, have been proposed in support of the Global

Table 5.2 Metadata descriptors (elements) included in the Dublin Core (Weibel 1997); NASA's Directory Interchange Format (DIF Version 5.0a; NASA Goddard Space Flight Center 1997); the Global Information Locator Service/Government Information Locator Service (GILS, see Barton 1996) and the National Oceanic and Atmospheric Administration Program's National Environmental Data Referral Service (NEDRES) formats.

Element no.	Dublin core	DIF	GILS	NEDRES
1	Title	Directory entry identifier	Title	Accession number
2	Creator	Directory entry title	Originator	Title
3	Subject	Data set citation	Contributor	Abstract
4	Descriptions	Investigator	Date of publication	Data collection description
5	Publisher	Technical contact	Place of publication	Data centre processing description
6	Contributor	Discipline	Language of resource	Period of record
7	Date	Parameters	Abstract	Length of record
8	Type	Keywords	Controlled subject index	Geographic place names
9	Format	Sensor name	Subject terms uncontrolled	Geographic codes
10	Identifier	Source name	Spatial domain	Geographic grid locators
11	Source	Temporal coverage	Time period	Parameter matrix
12	Language	Data set progress	Availability	Descriptors
13	Relation	Spatial coverage	Sources of data	Contact address
14	Coverage	Location	Methodology	Availability conditions
15	Rights	Data resolution	Access constraints	Principal investigator(s)
16		Project	Use constraints	Programme sponsor, contract, project, or experiment name
17		Aggregation flag	Point of contact	Processing/collecting organization
18		Quality	Supplemental information	Publications
19		Access constraints	Purpose	Related records
20		Use constraints	Agency programme	Date entered/updated
21		Originating centre	Cross reference	Category codes
22		Data centre	Schedule number	
23		Catalogue link	Control identifier	
24		Storage medium	Original control identifier	
25		Distribution	Record source	
26		Multimedia sample	Language of record	
27		Reference	Date of last modification	
28		Summary	Record review date	
29		DIF author		
30		IDN code		
31		DIF revision date		
32		Future review date		
33		Science review date		

Climate Observing System (Miller *et al.* 1996). Special types of metadata associated with large statistical databases, including logical or arithmetic expressions (e.g. weighting, aggregation, error), data quality indicators and statistical summary data are discussed by McCarthy (1982). Ziskin and Chan (1997) emphasize the importance of including data statistical summaries in metadata to support content-based data selection from large scientific databases, such as those associated with the US National Aeronautics and Space Administration's Earth Observing System.

5.4 Software and resources

Guidelines for metadata structure and supporting technology (i.e. user friendly software for metadata generation and management) are being discussed and developed by numerous organizations, including software vendors, government agencies, the National Center for Ecological Analysis and Synthesis, the US Long Term Ecological Research Network, the Centre for Earth Observation (European Commission) and scientific societies. For example, as part of the National Biological Information Infrastructure (NBII) in the United States, the NBII MetaMaker was developed to support geospatial metadata generation in a format that conforms to Federal Geographic Data Committee (FGDC 1994, 1998) guidelines. Numerous other metadata tools conforming to FGDC guidelines or representing subsets of the Content Standards (i.e. 'lite' versions) have been developed to meet different institutional objectives.

One particularly promising approach to metadata management is embodied in Web-based metadata search and data retrieval systems. An example of this is a software tool entitled Mercury that was developed at the Oak Ridge National Laboratory Distributed Active Archive Center (Kanciruk *et al.* 1999). Mercury supports searches of metadata to identify data of interest and deliver those data to the user. To make data and metadata available, data providers make them 'visible' in an area on their computer and Mercury periodically harvests the metadata and automatically constructs an index and a relational database that subsequently reside at a central facility. Web-based metadata management programs like Mercury have several benefits, including control of 'data visibility' by the scientist, high levels of inherent automation and computer platform independence.

Web sites for current information on metadata generation tools (e.g. NBII MetaMaker), international and United States Federal Geographic Data Committee activities, ecological metadata standardization efforts and related materials can be found at several of the following World Wide Web sites and additional pointers located therein (e.g., http://www.megrin.org/, http://www.lternet.edu/, http://www.fgdc/gov/, http://www-eosdis.ornl.gov/, http://www.eurogi.org/, and http://www.dlib.org/ [D-lib Magazine]). Several general issues (e.g. availability, cost) should be considered before acquiring existing metadata generation tools or developing new tools tailored to a specific need. General issues as well as a more detailed discussion of metadata structure and relevant programming tools are presented below.

5.4.1 Decision-making issues

Utilizing existing metadata generation tools whenever possible will avoid the costs and personnel time associated with developing new tools. When choosing among various options it is important to consider whether the software meets the specified objectives (especially, metadata completeness), and whether it conforms to industry-wide or discipline-specific guidelines. In some cases, it may be necessary to use more than one metadata generation tool. For example, an institution's spatial data may be incorporated into a GIS vendor-supplied metadata program that conforms to Federal Geographic Data Committee (1994, 1998) standards and is well integrated into the GIS environment. Their water quality data, in contrast, may be incorporated into a specific metadata program that meets other requirements established by a state or federal funding agency.

Some of the most important metadata attributes (e.g. natural history observations) are often recorded and maintained in unstructured format in the field or laboratory using pencil and paper. These attributes may be critical for correct data interpretation and analysis. Field notes and other unstructured metadata can either be archived in paper files or later converted into digital format (e.g. scanning, transcription to text or word processing files). These types of unstructured metadata may be suitable for exchange with expert colleagues but are inadequate for electronic data set publication and sharing with the broader scientific community. Although existing or proposed metadata generation tools may fill most of a project's metadata needs, consideration should be given to how maps, field notes and other unstructured data will be archived and managed, as well as referenced in the metadata.

Ideally, data and metadata should be independent of hardware and software to the greatest extent possible (Conley & Brunt 1991). Proprietary data storage formats inevitably change over time or are replaced by new formats. Thus, the life span (long-term utility) of data and metadata may be shortened when data and/or metadata conform to a proprietary standard as opposed to a more generic industry-wide standard. One such proposed generic standard, the Warwick Framework, describes data 'containers' for interoperability across data and scientific domains (Lagoze *et al.* 1996); current efforts are focused on developing mechanisms to support the transition from conceptual framework to digital objects (Daniel & Lagoze 1997).

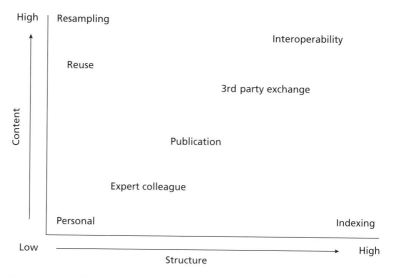

Fig. 5.2 Conceptual diagram illustrating the variable levels of content and structure that are necessary for meeting particular objectives. (Adapted from a figure made available by courtesy of James W. Brunt, LTER Network Office, University of New Mexico, New Mexico.)

5.4.2 Metadata structure

With high levels of secondary data usage and increases in metadata content, data utility may be improved by adding structure to the metadata (Fig. 5.2). Michener *et al.* (1997) described levels of metadata structure ranging from low to medium to high structure. A free-flowing narrative is an example of metadata with low structure. An example of a medium level of structure would be metadata consisting primarily of free-flowing narrative but including a small number of required, fixed-format fields that could support search and retrieval capabilities (e.g. by keywords, location, time period; see Conley & Brunt 1991; Ingersoll *et al.* 1997). A highly structured and fully searchable metadata record would include sophisticated database management system (DBMS) approaches. Minimal metadata structure should not be confused with low volume.

Increased metadata structure can be beneficial for several reasons (Michener *et al.* 1997). First, the checklist character of structured metadata often provides a memory-aid for the data originator about the information needed to facilitate subsequent data processing and interpretation. Second, increased structure facilitates development of searchable catalogues and database interfaces, potentially making the data available to a larger population of users and a wider range of processing software. High levels of structure may be good practice or, in some cases, may be required for specific projects (e.g. those requiring periodic data audits). However, highly structured metadata may be excessive where low levels of secondary usage are projected. Thus, the benefits of incorporating metadata into highly structured DBMS format should be considered in relation to software, programming, development and maintenance costs.

5.4.3 Tools for software development

The choice of metadata media and structure will often be dictated by availability of metadata generation tools and trained personnel, time and funding constraints, and projected rates of metadata usage. When specific metadata tools are inadequate or unavailable, metadata may be incorporated into word processing files (or free-flowing ASCII text), analytical programs (e.g. Statistical Analysis System programs), or more structured DBMS programs. Satisfying high levels of demand for metadata may necessitate making the metadata DBMS-accessible via the World Wide Web (see Chapter 3).

Many previous approaches to the management of ecological metadata focused on flexible formats, minimal structure, and implementation using ASCII text editors, word processing software and relatively simple relational DBMS (Gurtz 1986; Stafford *et al.* 1986; Michener *et al.* 1987; Conley & Brunt 1991; Brunt 1994; Briggs & Su 1994). During the mid 1990s, some organizations with moderate to large holdings of data began implementing metadata schemes utilizing format descriptions or more sophisticated DBMS software that increased metadata structure and often directly linked data and metadata. Primary objectives of these efforts were to initiate some standardization of content and structure to facilitate search and retrieval. For example, both the US National Oceanic and Atmospheric Administration's Earth System Data Directory (Barton 1995) and the US Geological Survey's Global Land Information System (Scholz & Smith 1995) utilize Directory Interchange Format (DIF) as a mechanism to ensure a minimum level of metadata is available during searches for data sets. Similarly, non-imagery data and associated metadata collected under the auspices of the US Department of Energy's Atmospheric Radiation Program are integrated and stored in Network Common Data Format (NetCDF) structure (Melton 1995). (For more information on NetCDF, see Rew and Davis 1990.)

Bretherton and Singley (1994) have identified a range of database functions that might be required in a scientific metadatabase management system. They also outlined an architecture for implementing these functions using existing DBMS and other software tools. Many organizations have already invested heavily (e.g. money, personnel, and training) in a specific DBMS, which could be modified for metadata management.

Several tools are available for implementing World Wide Web-based metadata applications, including Hypertext Markup Language (HTML) and eXtensible Markup Language (XML). XML is a standardized text format that represents a subset of Standard Generalized Markup Language (SGML; ISO standard 8879). Currently, XML offers the greatest utility as a representation language for documenting the content and semantics of Web-based resources. It was specifically designed for transmitting structured data to Web applications, and its utility is furthered by its relative ease of extensibility, a flexible structure that supports arbitrary nesting, and the potential for automated validation (Carlson 1997; Khare & Rifkin 1997).

5.5 Metadata implementation

Objectives for metadata implementation are straightforward, and include facilitating identification and acquisition of data for a specific theme, time period and/or geographical location; determination of the data's suitability for meeting a specific objective; and data processing, analysis and modelling. Effective strategies for metadata implementation may be less obvious because metadata encompass a diverse array of facts which are often not recorded in any systematic way and may only reside within the researcher's mind. At least three major issues warrant consideration during metadata planning and implementation: desired data longevity, projected rate of use and sharing of responsibility. In addition, several recommended 'keys to success' may facilitate the implementation process.

5.5.1 Planning issues

All data should be accompanied by at least a minimal set of metadata. The completeness of the metadata governs the length of time and the extent to which data can be reused by the original investigator(s) and utilized by scientists, resource managers, decision-makers and other potential users. Just as the data and information contained in a manuscript support peer-review of the publication and the conclusions reached therein, metadata support peer-review of the data and can facilitate secondary utilization. Although somewhat arbitrary, 20 years has often been cited as the objective for having data usable by scientists unfamiliar with the data and their collection (National Research Council 1991; Webster 1991; Strebel *et al.* 1994). Perceived data value, projected rates of secondary usage and financial and personnel costs should be considered in the context of establishing objectives for data longevity.

The development and maintenance of metadata can be a costly endeavour. Thus, it may be important to attempt to match the level of metadata content and format to the needs of anticipated users. Michener *et al.* (1997) identified three levels of secondary data utilization as representative of the relationship between metadata content and types of secondary usage. The first level is a colleague with technical expertise in the subject area and adequate knowledge of data collection, analytical and processing procedures. This user may require only data structural descriptors to make effective use of the data set. Second is an 'in-the-blind' user who discovered the data by means of searching an online metadata catalogue. In addition to data set and data structural descriptors, this individual is likely to require much more detail about research origins and data set status and accessibility. An ecologist who is attempting to reproduce computational results represents the third level. At this level, access to a very comprehensive set of metadata, including most or all supplemental descriptors and analytical programs, is needed. The required degree of sophistication of metadata content and format clearly varies across this range of potential uses.

Data reuse must frequently be based on intelligent and well-intentioned guesses. For example, the data originator (if still alive) may not remember what quality assurance procedures or analytical algorithms were used. If the relevant information was never documented or programmers or knowledgeable technical personnel have since left the project, a 'good guess' may be all that is available. Ultimately, the responsibility falls on both the data originator and secondary users to apply good practices and minimize the propagation of errors arising from unintentional 'misuse' of the data. For instance, the Carbon Dioxide Information Analysis Center (CDIAC; see Chapter 6) emphasizes the value-added component of data sets that results from the participation of scientists and users in metadata preparation, rigorous QA/QC processing, peer-review of data and metadata, beta testing of data sets prior to general release and incorporation of user feedback into its data packages (Boden 1995; NRC 1995).

Devoting resources during a project to metadata design and implementation costs money and personnel effort, and can result in fewer publications in the short-term. On the other hand, high quality data and metadata can be 'mined' for many years or even decades. Proper balance of short-term costs versus potential long-term gain is an issue warranting considerable thought and discussion by data originators, institutions, users and funding agencies.

5.5.2 Keys to successful metadata implementation

The first and probably most important component of metadata implementation is to perform a site or project needs assessment. Such an assessment can be conducted by a single individual (possibly a consultant) or by a small group of individuals that represent the spectrum of organizational or project interests. A needs assessment entails identifying data objectives (e.g. projected or desired data longevity, potential for reuse), establishing guidelines and procedures for data sharing and data ownership, assessing infrastructure (e.g. availability of hardware, software, personnel, funds), and categorizing and prioritizing metadata activities. For example, at a field station, meteorological data may receive a high priority for metadata implementation because of their perceived value to a large number of ongoing studies, historical usage patterns and potential for repeated use over time. In contrast, infrequent field surveys may receive a lower priority for long-term archival and metadata implementation. Once categories of data are prioritized, it is necessary to either adopt an existing metadata standard (e.g. geospatial metadata standard, FGDC 1994, 1998) or identify a set of minimal or optimal metadata descriptors that meet perceived needs.

A second recommended step in metadata implementation is to perform a pilot project using one or more relatively 'simple' data sets. Based upon successes and difficulties encountered in the pilot project, it is useful to re-evaluate site needs and objectives. For example, a formal or informal cost-benefit analysis may facilitate future prioritization and balance completeness of metadata versus funding and personnel availability. Following this evaluation process, it is often beneficial to formalize metadata activities. It may be desirable, for example, to develop relevant policies and procedures, to identify available metadata tools or initiate programming efforts to develop appropriate tools, and to establish a reward structure for providing comprehensive metadata. Metadata and other data management activities should be re-evaluated on a periodic basis to insure that they are meeting specified objectives.

Until formal metadata standards emerge and supporting software is developed, the metadata descriptors included in Table 5.1 may serve as the basis for initially developing metadata for individual scientists, laboratories and projects. Small groups of scientists focused on a specific research objective, such as synthesizing data on a particular topic, may benefit significantly from efforts to implement metadata. Compiling metadata for such a project would be an extremely daunting challenge for any one individual; however, workshops could provide a means by which scientists could exchange metadata, review metadata for completeness and comprehensibility, and fill in missing gaps (also see Hackos *et al.* 1997; Ingersoll *et al.* 1997). Successful completion of such comprehensive databases, including both data and metadata, may lead to new and innovative experiments and analyses. Important 'rules of thumb' that may facilitate successful implementation are reiterated in Table 5.3.

5.5.3 Integration, modelling and synthesis

Integration, modelling and synthesis projects are often hindered by the lack of high quality data and metadata (Rothenberg & Kameny 1994). Data, statistics and other parameters are routinely extracted from publications; unfortunately, publications often do not provide sufficient information pertaining to

Table 5.3 Metadata implementation 'keys to success'.

- Keep it simple! Start small and build upon successes. For example, the time and effort expended on a pilot project are usually paid back several-fold in the long run.
- Build consensus among scientists and data managers from the start. Data management initiatives, regardless of their potential benefits, are often unsuccessful when the 'user community' is excluded from the process.
- Data longevity is roughly proportional to metadata comprehensiveness. However, establishing a goal of **complete** metadata that can meet all future needs may be exorbitantly expensive and, ultimately, unattainable.
- Data and metadata should ideally be platform independent. Hardware and software change frequently. Today's 'standard' may be gone tomorrow. Thus, it pays to avoid proprietary storage formats whenever possible.
- The degree to which high-quality ecological data and accompanying metadata are securely archived and accessible for future research is directly related to the extent to which an ethic of data stewardship is promoted and rewarded.

the data distribution, requiring many assumptions about data ranges, frequency distributions, percentiles, etc. Ideally, raw data as well as the metadata that are critical for describing data collection objectives and methods, scale relevance of the data, and other potential limitations for secondary usage would be available for the synthesis activity. Metadata (i.e. station history files) have been integral to the data adjustments performed on US Historical Climatology Network data sets to improve data homogeneity (Easterling *et al.* 1996). The lack of such metadata for the Global Historical Climatology Network has necessitated the use of strictly nonmetadata-based adjustments.

Additional data and metadata limitations will become acutely obvious when ecological organizations attempt to develop data 'warehouses' to manage their institutional data resources. Data warehouses represent integrated, subject-oriented collections of data that have, thus far, been used primarily by the business sector (also see Chapter 7). The role of metadata in data warehouse implementation has been discussed by Inmon (1992), Grippo and Chen (1996), Kimball (1996) and Chaudhuri and Dayal (1997).

Some data sets, particularly long time-series, are associated with hypotheses involving complicated nonlinear relationships that are best represented by complex simulation models. Thus, preservation of the information about a set of data may also involve the preservation of the simulation model and its associated input and output files (Kirchner 1994). Peer-reviewed publications featuring simulation models tend to focus on the results and the conceptual and mathematical foundations for the model. Because simulation models tend to be modified through time, preservation of the model code and input files is likely to be critical if model experiments are ever to be truly reproducible. New information about the data set may come to light during the course of model execution, hypothesis testing and model validation; information that could benefit other secondary users. It may be useful to incorporate this information as well as a listing of publications resulting from secondary use of the data set into the metadata to facilitate additional secondary usage. Consequently, metadata related to version management, history management and derived databases warrant significant attention (Lanter 1991, 1993; Cammarata *et al.* 1994; Kirchner 1994; Rothenberg & Kameny 1994).

5.6 Metadata: challenges and opportunities

Basic and applied ecological research depend upon the availability of data as well as the ability to locate and use data. If a priori consideration is paid to the development of high quality data sets and accompanying metadata, then individual scientists and organizations can focus valuable time and effort on performing appropriate analyses. Individual scientists and organizations can benefit further from the availability of comprehensive metadata by being able to easily reuse data developed for other applications. Flexible metadata generation and management tools that support entry, search and retrieval are essential for facilitating metadata implementation. There is a significant need for research and development in this area.

Much future effort will focus on standardizing metadata content, accessibility, and structure (Fernandez-Falcon *et al.* 1993; Newton 1996; Stitt & Nyquist 1996; Ambur 1997; Gardner 1997). Where metadata standards are incomplete or do not exist, attention will likely focus on developing, endorsing and adopting appropriate standards (see Newton 1996 and Renner *et al.* 1996 for contrasting views on this subject). However, questions remain: who decides on the standard? and what constitutes minimal and optimal criteria (Drewry *et al.* 1997)?

Although a lack of standards may impede integrative research, Porter *et al.* (1997) caution that ecologists may likewise be constrained by the premature imposition of metadata standards before the science has advanced sufficiently. Consequently, they recommend the formalization of emergent standards, such as is occurring within the US Long Term Ecological Research Program (also see Ingersoll *et al.* 1997). Similarly, the Consortium for International Earth Science Information Network (CIESIN) found that a single uniform metadata content standard (i.e. DIF) did not satisfy objectives of their partner organizations and was not easily adapted (Bourdeau *et al.* 1997). CIESIN, therefore, adopted and modified several different standards that met institutional objectives, and provided various incentives (e.g. training programmes, supporting documentation and user manuals and technical support) to facilitate standards compliance and metadata quality. These and other standardization efforts (e.g. Australia New Zealand Land Information Council Metadata Guidelines, Shelley *et al.* 1997) indicate that the true test of any emerging metadata standard will be whether the standard is simple to use and understand, and whether it improves our science. Foresman *et al.* (1996) contend that the 'striking lack of compliance and implementation' of the FGDC Metadata Content Standards at local, state and federal levels is attributable to the standards being unnecessarily lengthy (> 200 metadata fields), tedious, convoluted and extremely costly to implement.

Funding agencies, scientific societies and research institutions should recognize that there are costs, as well as benefits, associated with archiving data and developing and maintaining the requisite metadata. Although future research endeavours will inevitably pay more attention to metadata and other aspects of data management, time is running out on many extremely valuable long-term and unique ecological data sets. There is a significant need for established funding mechanisms and data archives to support metadata development and secure long-term storage of these irreplaceable data (Gross *et al.* 1995). In addition, development of attendant reward systems (e.g. peer-reviewed data and metadata publications, equating database construction with publication efforts) will be essential for furthering an ethic of data stewardship (Porter & Callahan 1994; Callahan *et al.* 1996).

5.7 References

Ambur, O.D. (1997) Metadata or malfeasance: which will it be? In: *Proceedings of the Second IEEE Metadata Conference*. IEEE Computer Society (http://www.computer.org/conferen/proceed/meta97), Silver Spring, MD.

Barton, G.S. (1995) Directory Interchange Format: a metadata tool for the NOAA Earth System Data Directory. In: *The Role of Metadata in Managing Large Environmental Science Datasets Proceedings* (eds R.B. Melton, D.M. DeVaney & J.C. French), pp. 19–23. Pacific Northwest Laboratory, Richland, WA.

Barton, G.S. (1996) Multiple metadata formats from the NOAA Environmental Services Data Directory. In: *Proceedings of the First IEEE Metadata Conference April 16–18, 1996*. IEEE Computer Society (http://www.computer.org/conferen/proceed/meta96), Silver Spring, MD.

Boden, T.A. (1995) Metadata compiled and distributed by the Carbon Dioxide Information Analysis Center for global climate change and greenhouse gas-related data bases. In: *The Role of Metadata in Managing Large Environmental Science Datasets.* (eds R.B. Melton, D.M. DeVaney & J.C. French), pp. 13–18. Pacific Northwest Laboratory, Richland, WA.

Boone, R.D., Grigal, D.F., Sollins, P., Ahrens, R.J. & Armstrong, D.E. (1999) Soil sampling, preparation, archiving and quality control. In: *Standard Soil Methods for Long-term Ecological Research* (eds G.P. Robertson, D.C. Coleman, C.S. Bledsoe & P. Sollins), pp. 3–28. Oxford University Press, New York.

Bourdeau, R.H., Burley, C.J., Roseberry, G.S. & Sullivan, K.C. (1997) Institutions, languages, content standards, record syntaxes, and protocols: fitting it all together. In: *Proceedings of the Second IEEE Metadata Conference*. IEEE Computer Society (http://www.computer.org/conferen/proceed/meta97), Silver Spring, MD.

Bowser, C.J. (1986) Historic data sets: lessons from the past, lessons for the future. In: *Research Data Management in the Ecological Sciences.* (ed. W.K. Michener), pp. 155–179. University of South Carolina Press, Columbia, SC.

Bretherton, F.P. & Singley, P.T. (1994) Metadata: a user's view. In: *Proceedings of the*

Seventh International Working Conference on Scientific and Statistical Database Management. (eds J.C. French & H. Hinterberger), pp. 166–174. IEEE Computer Society Press, Washington, DC.

Briggs, J.M. & Su, H. (1994) Development and refinement of the Konza Prairie LTER research information management program. In: *Environmental Information Management and Analysis: Ecosystem to Global Scales.* (eds W.K. Michener, J.W. Brunt & S.G. Stafford), pp. 87–100. Taylor and Francis, Ltd., London.

Brunt, J.W. (1994) Research data management in ecology: a practical approach for long-term projects. In: *Proceedings of the Seventh International Working Conference on Scientific and Statistical Database Management.* (eds J.C. French & H. Hinterberger), pp. 272–275. IEEE Computer Society Press, Washington, DC.

Callahan, S., Johnson, D. & Shelley, P. (1996) Dataset publishing—a means to motivate metadata entry. In: *Proceedings of the First IEEE Metadata Conference.* IEEE Computer Society (http://www.computer.org/conferen/proceed/meta96), Silver Spring, MD.

Cammarata, S., Kameny, I., Lender, J. & Replogle, C. (1994) A metadata management system to support data interoperability, reuse, and sharing. *Journal of Database Management* **5**, 30–40.

Carlson, D. (1997) XML documents can fit OO apps. *Object Magazine* **7**, 24–26.

Chaudhuri, S. & Dayal, U. (1997) An overview of data warehousing and OLAP technology. *Sigmod Record* **26**, 65–74.

Conley, W. & Brunt, J.W. (1991) An institute for theoretical ecology? Part V: Practical data management for cross-site analysis and synthesis of ecological information. *Coenoses* **6**, 173–180.

Daniel, R. Jr. & Lagoze, C. (1997) Extending the Warwick Framework: from metadata containers to digital objects. *D-lib Magazine* **November 1997**. (http://www.dlib.org).

Defense Mapping Agency. (1992) *Vector Product Format, Military Standard 600006.* Department of Defense, Washington, DC.

Desai, B.C. (1997) Supporting discovery in virtual libraries. *Journal of the American Society of Information Science* **48**, 190–204.

Digital Geographic Information Working Group. (1991) *DIGEST: A Digital Geographic Exchange Standard.* Defense Mapping Agency, Washington, DC.

Drewry, M., Conover, H., McCoy, S. & Graves, S.J. (1997) Metadata: quality versus quantity. In: *Proceedings of the Second IEEE Metadata Conference.* IEEE Computer Society (http://www.computer.org/conferen/proceed/meta97), Silver Spring, MD.

Easterling, D.R., Peterson, T.C. & Karl, T.R. (1996) On the development and use of homogenized climate databases. *Journal of Climate* **9**, 1429–1434.

Federal Geographic Data Committee. (1994) *Content Standards for Digital Geospatial Metadata (June 8).* Federal Geographic Data Committee, Washington, DC.

Federal Geographic Data Committee. (1998) *FGDC-STD-001-1998. Content Standard for Digital Geospatial Metadata (revised June 1998).* Federal Geographic Data Committee, Washington, DC.

Fernandez-Falcon, E.A., Strittholt, J.R., Alobaida, A.I., Schmidley, R.W., Bossler, J.D. & Ramirez, J.R. (1993) A review of digital geographical information standards for the state/local user. *Journal of the Urban and Regional Information Systems Association* **5**, 21–27.

Foresman, T.W., Wiggins, H.V. & Porter, D.L. (1996) Metadata myth: misunderstanding the implications of federal metadata standards. In: *Proceedings of the First IEEE Metadata Conference.* IEEE Computer Society (http://www.computer.org/conferen/proceed/meta96), Silver Spring, MD.

Gardner, S.R. (1997) The quest to standardize metadata. *Byte* **22**, 47–48.

Grippo, K. & Chen, J. (1996) Building a data warehouse with SAS® software: a recipe for success. In: *Proceedings of the Twenty-First Annual SAS® Users Group International Conference*, pp. 686–692. SAS Institute, Inc. Cary, NC.

Gross, K.L., Pake, C.E., Allen, E., Bledsoe, C., Colwell, R., Dayton, P., Dethier, M., Helly, J., Holt, R., Morin, N., Michener, W.K., Pickett, S.T.A. & Stafford, S. (1995) *Final Report of* the *Ecological Society of America Committee on the Future of Long-term Ecological Data (FLED), Volume I: Text of the Report.* (http://www.sdsc.edu/~ESA/FLED/FLED.html).

Gurtz, M.E. (1986) Development of a research data management system: factors to consider. In: *Research Data Management in the Ecological Sciences.* (ed. W.K. Michener), pp. 23–28. University of South Carolina Press, Columbia, SC.

Hackos, J.T., Hammar, M. & Elser, A. (1997) Customer partnering: data gathering for complex online documentation. *IEEE Transactions on Professional Communication* **40**, 102–110.

Heery, R. (1996) Review of metadata formats. *Program: Automated Library and Information Systems* **30**, 4, 345–373.

Ingersoll, R.C., Seastedt, T.R. & Hartman, M. (1997) A model information management system for ecological research. *BioScience* **47**, 310–316.

Inmon, W.H. (1992) *Building the Data Warehouse.* John Wiley & Sons, Inc., New York.

Kanciruk, P., Gentry, M. & Rhyne, T. (1999) MERCURY—A Web-based metadata search and data retrieval system. In: *EOGEO99: Earth Observation (EO) & Geo-Spatial (GEO) Web and Internet Workshop '99.* Committee on Earth Observation Satellites. (http://webtech.ceos.org/).

Kellogg Biological Station. (1982) *Data Management at Biological Field Stations, Report of a Workshop Held May 17–20, 1982.* W.K. Kellogg Biological Station, Michigan State University, Hickory Corners, MI.

Khare, R. & Rifkin, A. (1997) XML: a door to automated web applications. *IEEE Internet Computing* **1**, 78–87.

Kimball, R. (1996) *The Data Warehouse Toolkit.* John Wiley & Sons, Inc., New York.

Kirchner, T.B. (1994) Data management and simulation modelling. In: *Environmental Information Management and Analysis: Ecosystem to Global Scales.* (eds. W.K. Michener, J.W. Brunt & S.G. Stafford), pp. 357–375. Taylor and Francis, Ltd., London.

Kirchner, T., Chinn, H., Henshaw, D. & Porter, J. (1995) Documentation standards for data exchange. In: *Proceedings of the 1994 LTER Data Management Workshop.* (eds R. Ingersoll & J. Brunt), pp. 5–8. Long-Term Ecological Research Network Office, University of Washington, Seattle, WA.

Lagoze, G., Lynch, C.A. & Daniel, R. Jr. (1996) The Warwick Framework: A container architecture for aggregating sets of metadata. *Cornell Computer Science Technical Report TR96–1593, July 1996.* (http://cs-tr.cc.cornell.edu/Dienst/UI/2.0/Describe/ncstrl.cornell/TR96–1593).

Lanter, D.P. (1991) Design of a lineage-based meta-database for GIS. *Cartography and Geographic Information Systems* **18**, 255–261.

Lanter, D.P. (1993) A lineage meta-database approach toward spatial analytic database organization. *Cartography and Geographic Information Systems* **20**, 112–121.

Leigh, R.A. & Johnston, A.E. (1994) *Long-term Experiments in Agricultural and Ecological Sciences.* CAB International, Wallingford, Oxon UK.

McCarthy, J.L. (1982) Metadata management for large statistical databases. In: *Proceedings of the International Conference on Very Large Databases*, pp. 234–243. Institute of Electrical and Electronics Engineers, New York.

Melton, R.B. (1995) Metadata in the atmospheric radiation measurement program. In: *The Role of Metadata in Managing Large Environmental Science Datasets.* (eds R.B. Melton, D.M. DeVaney & J.C. French), pp. 9–12. Pacific Northwest Laboratory, Richland, WA.

Michener, W.K., Feller, R.J. & Edwards, D.G. (1987) Development, management, and analysis of a long-term ecological research information base: example for marine macrobenthos. In: *New Approaches to Monitoring Aquatic Ecosystems. ASTM STP 940.* (ed. T.P. Boyle), pp. 173–188. American Society for Testing and Materials, Philadelphia, PA.

Michener, W.K., Brunt, J.W., Helly, J., Kirchner, T.B. & Stafford, S.G. (1997) Non-geospatial metadata for the ecological sciences. *Ecological Applications* **7**, 330–342.

Miller, C., Karl, T., Maiden, M. & Koltermann, P. (1996) Documenting climatological data sets for GCOS: a conceptual model. In: *Proceedings of the First IEEE Metadata Conference.*

IEEE Computer Society (http://www.computer.org/conferen/proceed/meta96), Silver Spring, MD.

NASA Goddard Space Flight Center. (1993) *Directory Interchange Format Manual, Version 4.1*. World Data Center—A for Rockets and Satellites, National Aeronautics and Space Administration, Goddard Space Flight Center, Greenbelt, MD.

National Geospatial Data Framework. (1998) *Discovery Metadata Guidelines*. National Geospatial Data Framework Management Board (http://www.ngdf.org.uk), Southampton, UK.

National Institute of Standards and Technology. (1992) *Spatial Data Transfer Standard (Federal Information Processing Standard 173)*. National Institute of Standards and Technology, Gaithersburg, MD.

National Research Council. (1991) *Solving the Global Change Puzzle, A U.S. Strategy for Managing Data and Information*. Report by the Committee on Geophysical Data Commission on Biosciences, Environment, and Resources. National Academy Press, Washington, DC.

National Research Council. (1995) *Finding the Forest in the Trees*. National Academy Press, Washington, DC.

Newton, J. (1996) Application of metadata standards. In: *Proceedings of the First IEEE Metadata Conference*. IEEE Computer Society (http://www.computer.org/conferen/meta96/meta_home.html), Silver Spring, MD.

Porter, J.H. & Callahan, J.T. (1994) Circumventing a dilemma: historical approaches to data sharing in ecological research. In: *Environmental Information Management and Analysis: Ecosystem to Global Scales*. (eds W.K. Michener, J.W. Brunt & S.G. Stafford), pp. 193–202. Taylor and Francis, Ltd., London.

Porter, J.H., Henshaw, D.L. & Stafford, S.G. (1997) Research metadata in Long-Term Ecological Research (LTER). In: *Proceedings of the Second IEEE Metadata Conference*. IEEE Computer Society (http://www.computer.org/conferen/proceed/meta97), Silver Spring, MD.

Renner, S.A., Rosenthal, A.S. & Scarano, J.G. (1996) Data interoperability: standardization or mediation. In: *Proceedings of the First IEEE Metadata Conference*. IEEE Computer Society (http://www.computer.org/conferen/meta96/meta_home.html), Silver Spring, MD.

Rew, R.K. & Davis, G.P. (1990) NetCDF: an interface for scientific data access. *IEEE Computer Graphics and Applications* **10**, 76–82.

Robertson, G.P., Coleman, D.C., Bledsoe, C.S. & Sollins, P. (1999) *Standard Soil Methods for Long-Term Ecological Research*. Oxford University Press, New York.

Rothenberg, J. & Kameny, I. (1994) Data verification, validation, and certification to improve the quality of data used in modeling. In: *Proceedings of the 1994 Summer Computer Simulation Conference*, pp. 639–644. The Society for Computer Simulation International, San Diego, CA.

Scholz, D.K. & Smith, T.B. (1995) The Global Land Information System: the use of metadata on three levels. In: *The Role of Metadata in Managing Large Environmental Science Datasets*. (eds R.B. Melton, D.M. DeVaney & J.C. French), pp. 25–27. Pacific Northwest Laboratory, Richland, WA.

Shelley, E.P., Johnson, B.D., Taylor, M. & Golding, P.J. (1997) Spatial metadata in Australia—a learning experience. In: *Proceedings of the Second IEEE Metadata Conference*. IEEE Computer Society (http://www.computer.org/conferen/proceed/meta97), Silver Spring, MD.

Stafford, S.G., Alabach, P.B., Waddell, K.L. & Slagle, R.L. (1986) Data management procedures in ecological research. In: *Research Data Management in the Ecological Sciences*. (ed. W.K. Michener), pp. 93–114. University of South Carolina Press, Columbia, SC.

Stitt, S. & Nyquist, M. (1996) Development of a metadata content standard for biological resource data: National Biological Information Infrastructure draft metadata standard. In: *Proceedings of the First IEEE Metadata Conference*. IEEE Computer Society (http://www.computer.org/conferen/meta96/meta_home.html), Silver Spring, MD.

Strebel, D.E., Meeson, B.W. & Nelson, A.K. (1994) Scientific information systems: a conceptual framework. In: *Environmental Information Management and Analysis: Ecosystem to Global*

Scales (eds W.K. Michener, J.W. Brunt & S.G. Stafford), pp. 59–85. Taylor and Francis, Ltd., London.

Taylor, L.R. (1989) Objective and experiment in long-term research. In: *Long-Term Studies in Ecology* (ed. G.E. Likens), pp. 20–70. Springer-Verlag, New York.

Thiele, H. (1998) The Dublin Core and Warwick Framework. *D-lib Magazine* (January) (http://www.dlib.org).

Webster, F. (1991) *Solving the Global Change Puzzle: A US Strategy for Managing Data and Information*. Report by the Committee on Geophysical Data Commission on Geosciences, Environment and Resources, National Research Council, National Academy Press, Washington, DC.

Weibel, S. (1997) The Dublin Core: a simple content description model for electronic resources. *Bulletin of the American Society for Information Sciences* **October/November**, 9–11.

Ziskin, D.C. & Chan, P. (1997) Innovations in response to floods of data. In: *1997 International Geoscience and Remote Sensing Symposium*, pp. 1255–1256. Institute of Electrical and Electronics Engineers, New York.

CHAPTER 6

Archiving Ecological Data and Information

RICHARD J. OLSON AND RAYMOND A. MCCORD

6.1 Introduction

Traditionally, an archive is a place where documents and other materials of public or historical significance are preserved. Similarly, a data archive is a collection of data sets, usually electronic with accompanying metadata, stored in such a way that a variety of users can locate, acquire, understand and use the data. In addition, data in an archive are secure against natural and man-made disasters, and are preserved in a form that will continue to be accessible as technology changes. Most large research projects, government agencies, non-governmental organizations and private foundations involved in ecology store their data holdings in keeping with locally defined protocols and procedures, which may or may not satisfy the requirements of an archive. Meanwhile, over the past 30 years, ecology has been shifting from traditional studies of isolated populations, communities and ecosystems to more broad-scale approaches involving modelling, synthesis and assessment studies (Fig. 6.1). As ecology moves toward regional and global multi-disciplinary studies, mechanisms are needed to promote sharing of data among many disciplines (meteorology, hydrology, soil science, forestry, agriculture, botany, etc.) and to insure future data accessibility (NRC 1995a, 1997). Formal archives for ecological data will someday make this process possible. Submitting data to archives and acquiring data from archives will become an integral part of tomorrow's scientific process. However, archiving of data is not yet given the attention, resources, or recognition required to become a routine part of the research and publication cycle (NRC 1997). Although there are a limited number of formal data archives for ecological data, ecologists can manage their data today in ways that meet local needs for security, facilitate data sharing and will prepare them for eventual inclusion in data archives. This includes having an understanding of how data should be prepared for archiving, the components and functions of data archives and the future directions for data archives.

6.1.1 What is a data archive?

A data archive is established for the 'preservation with understanding' of ecological data. Optimally, it is a permanent collection of data sets with accompanying

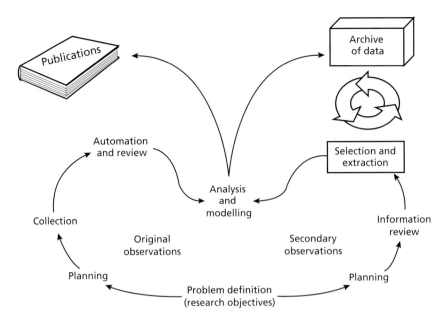

Fig. 6.1 The information perspective of scientific research follows two pathways: (a) original observations; a pathway with hundreds of years of experience and development; (b) secondary observations from data reuse; a pathway with 20–30 years of experience and significant development needed.

metadata such that a variety of users can readily acquire, understand, and use the data. The design of an archive is defined by three constraints:
1 scope of the data to be stored;
2 capabilities for searching and access;
3 resources for operation and maintenance (O'Neill *et al.* 1993).

A data archive preserves data and metadata in a form that will continue to be accessible as technology changes. Equally important is that it provides complete metadata so data users will understand any inherent limitations associated with using the archived data in new applications. Whereas 'archive' in its common usage in ecology may imply simple preservation, a worthier goal is facilitating data sharing and fostering broader ecological discoveries. Archives can be more than a long-term backup, and more than an index or catalogue with pointers to data sets stored elsewhere.

As indicated in Table 6.1, different terms and functions are often associated with data sharing and storage activities. Informal sharing, single project data management systems, repositories and digital libraries may provide some or all of the functionality of an archive. The concepts presented here for preparing data for archives apply readily to these less formal activities.

Table 6.1 Distinctions between data exchange and storage activities.

Venue	Data exchange	Comments
Data custodianship	Data sharing by request, usually with colleague having equivalent technical expertise	Current expert, provides technical information, could authorize changes to the data, may be the primary compiler or immediate heir to the data
Data stewardship	Data sharing by request, may have limited access	Gatekeeper, minimal knowledge of the data, may inherit data from custodian, no advertising of the data availability
Data repository	May have limited access	Usually a project-level function; limited access, limited functionality
Digital library	Public access	Broad subject area, limited expertise and user support, includes tabular and graphic data
Data catalogue or Clearinghouse	Public access	Broad subject area, maintains searchable descriptors without data, provides index and pointers to data
Data archive	Public access	May have thematic emphasis, search and order, long-term commitment, packaged metadata

6.1.2 Formal data archives

Data archives have been more fully developed for environmental information that has a limited theme and is a collection of large amounts of data over time and space. The US National Oceanographic and Atmospheric Administration's (NOAA) National Climatic Data Center (NCDC) is a good example. Not only is it the longest operating data centre in the US (established in 1951), but it has the longest period of data (since the 1800s), has the largest environmental archive in terms of data volume (55 gigabytes added daily), and has the largest number of users (over 170 000 requests annually) (see www.ncdc.noaa.gov/ol/about/whatisncdc.html). NCDC provides the historical perspective on climate, an essential component of ecological research, particularly for global climate change studies. Under its authorizing legislation, the NCDC is designated as the US Department of Commerce's Agency Records Center and it provides climate data in over 15 possible forms. Paper copies of many of the weather records are submitted to the National Archives for permanent storage. The Center also operates the World Data Center-A for Meteorology, which provides for international data exchange. NCDC provides a comprehensive model of how to archive climate data.

Two additional examples of comprehensive archives of ecological data are the Carbon Dioxide Information Analysis Center (CDIAC) and the Oak

Table 6.2 Examples of data archives.

Data archive & Web address	Focus	Special features
National Climatic Data Center (NCDC) www.ncdc.noaa.gov/	Acquire, distribute and archive climatic data	A very long-term collection of a limited number of data types from many locations
Carbon Dioxide Information Analysis Center (CDIAC) cdiac.esd.ornl.gov	Acquire, compile, quality-assure, document, archive and distribute information on greenhouse gases and climate change in support of the US Department of Energy's Global Change Research Program	Special emphasis on quality assurance, documentation and derived, integrated products. User community of many thousands of researchers, policy makers, educators, students and corporate officials around the world
Distributed Active Archive Center for Biogeochemical Dynamics (ORNL DAAC) www-eosdis.ornl.gov	Acquire, document, quality-assure and archive multi-disciplinary data for terrestrial ecosystems and provide access to the global change research community, policy makers and educators. Part of the National Aeronautics and Space Administration's (NASA) Earth Observing System (EOS)	Web-based data search and order interface, browse and view data before ordering, multiple distribution media, free and ready access, User Services Office

Ridge National Laboratory Distributed Active Archive Center (DAAC) for Biogeochemical Dynamics (Table 6.2). CDIAC, which is funded by the US Department of Energy, has a reputation for distributing data of high quality with known accuracy and reliability (NRC 1995b). It has produced data packages since 1982 that include extensive data description, detailed format information, extensive quality assurance (QA) checks, examples of data applications and copies of key associated literature. CDIAC produces value-added products for the global change research community and uses a process by which data are certified as valid by those that collected the data and carried out the QA checks (NRC 1997).

The Oak Ridge National Laboratory's (ORNL) Distributed Active Archive Center (DAAC) for Biogeochemical Dynamics is operated as part of the US National Aeronautics and Space Administration's (NASA) Earth Observing System (EOS). EOS consists of a science component and a data system supporting a coordinated series of satellites for long-term global observations. The Data and Information System component (EOSDIS) is a distributed system to support archiving and distribution of data at multiple data centres or DAACs. These centres are connected by an information management system that provides an interface for 'one stop shopping' for earth science data, allowing users to search for and order data from multiple data centres in a single session.

Although the DAACs primarily archive imagery data from the EOS satellites, the ORNL DAAC maintains data related to biogeochemical dynamics to support the validation and interpretation of satellite products. The ORNL DAAC stores a relatively small portion of the EOS-generated data; however, it is building on the large investment that NASA is devoting to data archiving.

Many other data facilities and research networks store and distribute environmental and ecological data, but represent different stages of development toward a comprehensive data archive (Table 6.3). Many of these systems are evolving rapidly because of significant changes in technology such as Internet access and reduced costs for storage devices. These existing systems have typically contained a small number (10s) of measurement types generated by a few types of instruments, methods and sensors. Examples include the Long Term Ecological Research (LTER) network, which provides leadership in the organization and documentation of ecological data for archive purposes; the National Oceanic and Atmospheric Administration (NOAA) operates the US National Ocean Data Center, the US National Geophysical Data Center, and the US National Environmental Satellite, Data, and Information Service (NESDIS). The International Soil Reference and Information Center (ISRIC) uses a representative approach to storage of large-scale soils information. The US National Soils Data Access Facility (NSDAF) project is a multifaceted effort that will provide a mechanism for both internal and external customers to access, analyse, download and report the various national soils databases. The US Geological Survey operates the National Water Information System and maintains other collections of water resource information. Data archives have also been established for satellite information (e.g. NASA's EOSDIS DAAC). Data archives with a biological focus include the US Forest Service Forest Inventory and Analysis Database and the Breeding Bird Survey (Sauer *et al.* 1997). The Swedish Museum of Natural History has several databases containing ecological information. For instance, the database of the Linnean herbarium presents information on 1427 herbarium sheets collected during the 1800s including photos of original herbarium sheets and field notes. The museum databases also contain automated records describing 10 000s of fish specimens and collection notes.

Several data 'directories' provide information about available data (metadata) but do not store the actual data. The European Environmental Agency maintains a Catalogue of Data Sources (CDS). CDS is a locator system for environmental information. It supplies information on who has what information in Europe, in which form and where the data exist as well as how to gain access. In other words, the CDS provides meta-information to the users of environmental information and data, helping them to locate and retrieve relevant sources. Likewise, the US National Environmental Data Index (NEDI) provides direct access to environmental data and information descriptions and thereby improves awareness of and facilitates access to data and information

Table 6.3 Examples of environmental data storage and distribution activities. Data centres vary in their mode of operation, including broad versus narrow thematic scopes; pointers to distributed data holdings versus holding data and metadata; controlled versus public access; and short- versus long-term commitment to archive.

Data centre	Web URL	Agency[1]
Breeding Bird Survey (BBS)	http://www.mbr.nbs.gov/bbs/bbs.html	US FWS
Databases at the Swedish Museum of Natural History	http://www.nrm.se/databas.html.en	SMNH
Ecological Monitoring and Assessment Network in Canada	http://www.cciw.ca/eman-temp/tools/datasets/intro.html	EC
Environmental Information Centre	http://www.nmw.ac.uk/ITE/eic.html	NERC
European Topic Center on Catalogue of Data Sources	http://www.mu.niedersachsen.de/cds/	EEA
Forest Inventory and Analysis (FIA)	http://www.srsfia.usfs.msstate.edu/	USFS
Global Change Master Directory (GCMD)	http:..//gcmd.gsfc.nasa.gov/	NASA
Global Population Dynamics Database	http://cpbnts1.bio.ic.ac.uk/gpdd/	CPB, NSF
ISRIC Soil Information System	http://www.isric.nl/ISIS.html	ISRIC
Long Term Ecological Research Network (LTER)	http://lternet.edu/	NSF
National Environmental Data Index (NEDI)	http://www.nedi.gov/	NOAA
National Environmental Satellite, Data, and Information Service (NESDIS)	http://ns.noaa.gov/NESDIS/NESDIS_Home.html	NOAA
National Geophysical Data Center (NGDC)	http://www.ngdc.noaa.gov/	NOAA
National Ocean Data Center (NODC)	http://www.nodc.noaa.gov/	NOAA
National Soils Data Access Facility (NSDAF)	http://www.statlab.iastate.edu/soils/nsdaf/	NRCS
National Water Information System (NWIS)	http://waterdata.usgs.gov/nwis-w/US/	USGS
NERC Data Centres	http://www.nerc.ac.uk	NERC
Threatened and Endangered Species	http://www.fws.gov/~r9endspp/endspp.html	US FWS

[1] Agencies: CPB, Center for Population Biology; EC, Environment Canada; EEA, European Environmental Agency; ISRIC, International Soil Reference and Information Center; NASA, National Aeronautics and Space Administration; NERC, Natural Environment Research Council; NOAA, National Oceanic and Atmospheric Administration; NSF, National Science Foundation; NRCS, National Resource Conservation Service; SMNH, Swedish Museum of Natural History; US FWS, US Fish and Wildlife Service; USFS, US Forest Service; USGS, US Geologic Survey.

holdings. The overall goal of NEDI is to facilitate the use of the widest possible range of environmental data and information. NASA's Global Change Master Directory is a comprehensive directory of descriptions of data sets of relevance to global change research, including data sets covering climate change, the biosphere, hydrosphere and oceans, geology, geography and human dimensions of global change.

6.2 Factors affecting archiving of ecological data

Most ecologists support the concept of data archiving and even use data from archives in their research but, generally, they do not archive their own data. Most often, the user of archived data is perceived as deriving more benefits than the contributor (Porter & Callahan 1994). The Ecological Society of America's Future of Long-term Ecological Data (FLED) Committee identified seven factors associated with successful data banks and five concerns associated with data exchange networks (Gross *et al.* 1995). They recognized that long-term ecological data are being lost after a project is completed because investigators move on to other interests and the data are rarely archived. While there is a trend among federal funding agencies to require that data generated by supported projects be formally archived, the funding agencies generally attach a low priority to funding data management and archiving (NRC 1995c). Some of the factors contributing to avoidance of data archiving, some incentives promoting data archiving and levels of data sharing will be explored in this section.

6.2.1 Avoiding data archiving

The process for archiving data is equivalent to preparing a publication. The time and resources required to archive data can be equal to or exceed those for preparing, editing and reviewing a publication. However, the archival process is often poorly defined, with uncertainty about the detail, format and style of metadata (see Michener *et al.* 1997; Chapter 5). Unfortunately, the scientific community, especially administrative levels, does not acknowledge well-documented data as an equivalent accomplishment to a publication. Therefore, when resources are limited, scientists frequently emphasize publication writing and neglect preparation of data for archiving (Porter & Callahan 1994).

6.2.2 Incentives for data archiving

Project leaders, sponsors, and science managers can provide the following incentives for investigators to archive data:
1 establish a citation policy to give credit to data contributors and processors;
2 establish a citation policy to give credit to multiple contributors to integrated data sets and the integrator;
3 provide investigators with adequate resources, guidance and planning for data management;
4 involve data management staff in the initial project planning;

5 plan for 'value' in resulting data sets (one of the stated objectives of the project or programme should be the production of useful data);
6 provide guidelines and training for metadata preparation;
7 produce high visibility data products (e.g. CD-ROMs or hardcopy data products); and
8 give credit to those that produce well-documented data sets (i.e. include in promotion and salary actions).

These incentives would complement and supplement the incentives associated with the publication from the research.

On a larger scale, the Ecological Society of America (ESA) is attempting to facilitate free and open exchange of data among ecologists and to create the reward structure necessary to encourage ecologists to share and exchange data by supporting an electronic publication (*Ecological Archives*) that publishes peer-reviewed data sets (Peet 1998). The ESA Data Sharing and Archive Committee has developed guidelines for publishing data and metadata, as well as an appropriate peer-review process for 'data publishing'. Under the scheme, the data and metadata are maintained in a long-term archive and contributors and users are able to cite the published 'data papers', much like citing publications in other ESA journals (see http://esa.sdsc.edu/Archive/). Similarly, the British Ecological Society provides a venue for publishing ecological data as part of the *Journal of Ecology* data archives (http://www.demon.co.uk/bes/jearcmen.ht). This archive contains material in html files, such as data and complex methods that are referred to in papers printed in *Journal of Ecology*. In this case, authors are solely responsible for the accuracy of their archive material and no peer review is conducted.

6.3 Phases of data sharing

The flow of data is from an investigator who collects, processes, analyses and publishes data to other functions that enable the sharing, integrating, analysing and publishing of the data by others (Fig. 6.1). During this flow, the interest in and use of the data may change depending on the mechanism selected for sharing the data (Table 6.4). The maturation of data associated with large field projects is described by Strebel *et al.* (1994) as analogous to the progression from infancy through adolescence to maturity. According to Strebel *et al.* (1994), in the infant stage, data are collected and processed; in the adolescent stage, colleagues may provide peer review and the data may be revised. The adult stage is a free-standing, independent, published data set. A key factor is the avoidance of what Michener *et al.* (1997) describe as 'information entropy'. This term describes the long-term degradation of information content associated with data as an investigator's personal knowledge of the data is forgotten or data or metadata are accidentally lost (see Chapter 5). The role of archives in deterring this 'decay' process is described in this section. Data

Table 6.4 Phases of the data activity and availability.

	Phases of data availability		
Characteristic	Pre-published period	Post-published period	Extended period
Investigator involvement	high	high to low	low to none
Publication authorship	investigator and/or project authorship	investigator, colleagues, and/or project	synthesis studies, foreign investigators
Primary access	investigator	investigator, project DIS, archive	literature or archive
Users	investigator or colleagues	colleagues or multidiscipline investigators	multidiscipline investigators, 'science fair' studies
Required documentation	minimal	varies	complete
Access limitations	investigator permission; incomplete QA and processing, existence of data unknown	investigator permission, incomplete documentation	none
Use enhancements	provide beta release, involve users in QA, establish investigator-based Web pages	combine with collection of data, project Web page	submit to data archive or clearinghouse, advertise data availability, complete documentation

archiving is one way to promote the sharing of data, to prevent information entropy and to accumulate 'equity' from research investments.

6.3.1 Investigator's exclusive interest

Typically, the investigator has the greatest and most immediate interest in analysing and publishing the data he or she collected. Only limited data sharing may occur until these initial interests are satisfied. This can be especially true of graduate students striving to complete a dissertation and new investigators establishing a publication record. Responding to local and immediate pressures generally results in the data processing occurring before the metadata are either forgotten or recorded. Unless incentives are in place, providing metadata becomes a very low priority and may not occur. Efforts to archive data also fail when investigators fear that providing immediate access to their data may result in others 'scooping' the investigator's publication of his or her findings. This is a sensitive issue; however, this argument can be abused when investigators retain data for years without publishing. Obviously, such data will be subject to information entropy. Some funding agencies are providing the incentive for publication and archival by specifying an exclusive time period for investigators to publish, after which the data are to be made available. For instance, the National Science Foundation promotes timely availability of data from sponsored projects; some programmes like Long Term Ecological Research have established 2 years from time of collection as the time frame for making most data publicly available.

In multi-investigator projects, the project's milestone deadlines may supersede those of individual investigators, resulting in other data problems. When individual investigators are unable to provide their data to meet project-wide needs, these larger projects may be delayed. If the unpublished linkage to metadata is insufficient for the more general purposes of the data, the data are incomplete or there are inconsistencies in data management, a project-wide analysis or model development may be slowed. Sometimes investigators can share specific information or less sensitive data summaries while continuing to analyse their more detailed data. When scientists design data management and archiving as integral parts of their scientific process, delays associated with completing analyses and publications, as well as contributing data to archives, can be reduced.

6.3.2 Investigator to investigator sharing

New data are unique prior to publication. Co-investigators on a large project and colleagues are a potential group interested in newly collected data. Typically, this group knows the investigator, has equivalent technical expertise, and can recognize data limitations. Therefore, they may request the data directly from the investigator and do not require extensive documentation to use the data.

6.3.3 Open sharing with secondary users

Once the data are described in a publication, an additional set of users becomes interested in the data. Several factors may promote a high level of interest in and demand for a particular data set. As more publications cite a particular study and its data, there may be a positive feedback in user demand for the data. A particular data set may be unique or be the first in a new area of research (e.g. long-term CO_2 measurements by Keeling *et al.* 1982).

In the past, this group of data users may have resorted to extracting data from figures in publications or using published summary information. Because ecology is a multi-disciplinary science, this next generation of data users may be from disciplines very different from that of the original investigator. These users may not be as familiar with the data as a colleague and require more documentation to know the limitations associated with using the data in their particular application (Strebel *et al.* 1994). These users generally will acquire the data from an *archive* or as a stand-alone package from the scientist. By this time, the original scientist may have begun another project and may not have the resources or interest to provide additional technical support to these users. Research by these users would benefit greatly from the availability of data in a publicly accessible, comprehensive data archive.

If it requires months or years to get data into an archive for access by others, then the archiving process may become a constraint to conducting science involving synthesis and modelling. Modellers will use what's available, or generate estimates, when experimental data are unavailable. Some of the limitations and enhancements associated with the availability of data over the 'life stages' discussed previously are summarized in Table 6.4.

6.3.4 Data rescue

Sometimes data sets may enter an extended period (possibly several decades long) of low interest, followed by renewed interest when they become part of a larger collection (spatially or temporally larger). Long-term data are especially valuable for detecting rare events, subtle processes and complex phenomena in ecology (Gross *et al.* 1995). Often these data sets are 'orphaned' because the investigator and documentation may not be readily available, or the data were recorded in an electronic format (or on punched cards), with no current technology that can be used to read the old files. The challenge is to be able to identify potential orphan data with legacy value that may become a key to understanding ecological trends and patterns. For any data set not archived, or having inadequate documentation, this future opportunity for ecological research and knowledge is lost.

An example of a data compilation and rescue effort is the British Museum of Natural History hosting of the Web-based Caterpillar Hostplants Database (http://www.nhm.ac.uk/entomology/hostplants/index.html). This project offers a central facility to pool and permanently record rearing data for all Lepidoptera groups from anywhere in the world. Records of caterpillar host plants are scattered through published and manuscript sources and are difficult to retrieve. Many rearing records are never published and so are not accessible to other entomologists. However, collected host plant records in this database form a valuable scientific resource that can be used to answer broader biological questions about how Lepidoptera and plants interact. Another example of data rescue is the recent publishing of a collection of over 700 estimates of global terrestrial net primary productivity that were compiled in the early 1980s but were never made available to the scientific community. These data were reviewed, documented, re-organized, published (Esser *et al.* 1997) and distributed through the ORNL DAAC (http://www-eosdis.ornl.gov/npp/).

6.4 Preparing data for archiving

The limited development of ecological data archives results from the complexity and diversity of the data and the absence of broad-scale, long-term programmes (most research is multi-year). The complexities of ecological data and their almost unlimited structural diversity have been managed with one of the following approaches.

1 'Push' the data into a generic structure representing objects, attributes and time. The details of methods, locations, units, quality, etc. are managed in secondary dictionaries or lists. An integrated database results from this approach. An example is given by Moore (1997) who describes an information structure in which all measurements (scientific observations) can be put into a generic form that relates the object measured (unchanging attributes such as its name, type, etc.) with its features (measured and varying attributes such as position, size, physical or chemical activity, etc.), and time of the measurement. The generic form of structuring information is proposed to minimize the effort required for future data use.

2 Package each data set with independent, detailed documentation and index the collection of data sets (for example, see data packages prepared by CDIAC). The 'package' form of structuring information minimizes the effort required for the initial preparation of the data storage.

Both of these approaches accomplish the storage objective for data archival. However, finding and retrieving the data may be limited by the user's knowledge of the logical language used to describe the data and the user's patience to conduct thorough searches within the archive.

While data may be collected by a project having a plan and funding resources for the data to be deposited in a specific data archive (e.g. NASA's EOSDIS program), most data follow a more haphazard path to an archive. The data may be stored when the investigator moves to another place or project. The data may become part of a synthesis or modelling project with ties to an archive. Their fate is related to an individual's interest and determination to share data.

This section describes how an investigator should prepare a data set for submission to an archive (Table 6.5). The discussion emphasizes tabular data; however, reports, manuscripts, photographs, remotely sensed images, GIS coverages, audio and video clips can be submitted to an archive in an electronic form following a similar process.

6.4.1 Rules for good archive data

Based on authors' experience in compiling, managing and distributing ecological data since the early 1970s, some data and metadata problems are

Table 6.5 Scientist/data source functions and components.

Function	Component	Comments
Input	Create data dictionary	Variable names, labels, formats, units of measure, valid values, missing value codes,
	Format and file structure	Test the sufficiency of the structure to contain the data.
	Input data	Implement logs or records of input progress
Quality assurance	Verify input	Double entry, visual checks, check sums
	Validate	Check for data contamination, outliers and referential integrity
Document	Acquire guidelines	Michener *et al.* 1997, NASA EOSDIS Guides
	Input metadata	Check the quality of the input; metadata errors can be difficult to correct later
	Select descriptors	Search and order keywords, locators, etc.
Distribution	PI-to-PI data exchange	Optional: providing access to data

encountered too frequently. This experience has led to formulating the following guidelines for storing data in an archive. As with most situations, rules (even those that seem obvious and mundane) are reactions to past problems and deficiencies. Slagel (1994) has discussed the need to incorporate information standards as a fundamental step to achieve efficient data integration for environmental applications. The standards suggested may be compromised; however, the consequences of the compromises must be evaluated with respect to the final scientific utility of the data.

Data are organized to meet the needs of the current users. Often, investigators use spreadsheets or simple tabular approaches to enter and analyse their data. However, as data become incorporated into a project data and information system or archive, other database designs may be selected to be more flexible or efficient. Once the data are stored using a specific design, such as a relational database, the user may need to spend significant time to retrieve and reorganize the data.

Unique occurrences

Each type of measurement is represented in a consistent 'one way' in a data set, that is, the same units of measure, same method name, same analyte or parameter name throughout the data set. 'One way' means all occurrences of equivalent attributes or results have exactly identical representations; that is, no changes in the usage of case, spacing, punctuation, string length, etc. Each measurement event is represented by only one value in the data set: that is, no duplicates. Measurements are based on published methods. If the research includes method development, then the methods must be completely described in the publication. New methods must be clearly linked by code or name to the pertinent results.

Identifiers

Each value is associated with a parameter name in an unambiguous way, such as on the same record or having a relationship that cannot be easily altered. Each measurement value has a quality indicator or way to indicate or flag problems, and is associated with a unique event name (or sample ID).

Place and time

Each value is associated with a unique place name (or location ID) and each place name is given a quantitatively defined location (i.e. latitude and longitude or other coordinates). Event and place names should not be represented by coded information that is not stored in other ways. That is, do not create 'permanent' names subject to measurement error and to later revisions. (For example, place names should not be the coordinates of the place, because, if erroneous, they will be revised and make the names inconsistent; sample IDs should not include time.) Each value is associated with a date and time.

Data storage and transport

Data are stored or managed with a database management system or 'data-smart' software (e.g. S+, SAS), not with word processing or loosely defined spreadsheet formats. The enforcement of syntax and format rules as well as the structural logic makes word-processing and spreadsheet software insufficient and inappropriate for quality data. The logic in these formats is insufficient for the required QA reviews of the data. The structural integrity of the data and metadata is retained and described when the data are exported to 'vendor-neutral' data formats. These formats are required to assure the long-term accessibility of the data, which are documented by a data dictionary that defines field types, field definitions, lengths, valid values, code definitions, sequences and key fields linking related data sets (see Chapter 5 for more information). Data transported between organizations are accompanied by summary statistics to confirm a correct transfer, including number of records, and mean, minimum and maximum number of missing values for each parameter.

The 20-year rule proposed by the National Research Council (1991) is a practical guideline for archiving data and metadata. The rule states that 20 years after the data are archived, someone not familiar with the data or how they were obtained, should be able to understand and use the data based solely on the documentation (NRC 1991; also see Fig. 5.1).

6.5 Operating the data archive

The goal of an archive is to be able to use data in the future. In order to meet this goal, the archive must preserve the data, document the data to guide data use, provide a mechanism by which users can locate desired data, and transfer the data from the archive to a user.

To perform these basic functions, an archive must also be able to add new data to the archive, provide user support and maintain the computer system, including future upgrades of software, hardware, storage media and network connectivity. An archive optimally supports the basic data management system functions as described in Chapter 2, plus some special storage functions. This section describes each function as summarized in Table 6.6 and Fig. 6.2. Figure 6.2 shows how all of the functions relate to each other and emphasizes the core functions.

6.5.1 Adding data

Although there may be many variations on the overall process, an initial step for adding data to an archive is to connect a contributor with the appropriate data archive. Adding data is an ongoing process that requires identifying new needs of users and the availability of new data, and establishing priorities to acquire new data. Often a data archive is associated with a specific programme, or has a thematic orientation, and archive personnel will actively seek selected data sets. Guidelines for formatting and submitting data and metadata are a necessary part of an archive, enabling the contributor to prepare data and metadata as completely as possible for submission.

6.5.2 Data and metadata review

Archive staff review the data and metadata and may reformat them to achieve consistency and completeness. The quality assurance performed on the data as documented in the metadata is also reviewed. Archive staff and contributors work together to resolve questions, review changes and make sure the metadata supply citation information correctly acknowledging the original data contributors. After consensus is reached, data are entered into the archive database for public access and long-term archival storage (see below).

Table 6.6 Data archive functions and components.

Function	Components	Comments
Adding data	Define guidelines	Michener *et al.* 1997
	Confirm correct data transfer of data and correct formats	Counts of records, check statistics, check sums
	Review data/metadata	Evaluate its completeness and reasonableness
	Organize data for archive	1 collection of files or 2 enter data into RDBMS
	Ingest data	Enter metadata, search descriptors
Data preservation	Backup and security	Store copies of data in separate, secure location paper media option
	Storage systems	
	Migration to new media	Plans for migrating to new media (e.g. diskettes to CD-ROMs) and storage devices
Locating data	Index system	Follow keyword standards
	User interface	Determine if the data can be readily 'found' with the available logical options
	Subsetting	Provide tools to subset data and select parameters of interest
Data delivery	Media choices	Options: Web transfer, FTP transfer, CDROM, diskette, tapes, paper
System development	Hardware, software, storage media and network linkages	Allocate resources (staff, equipment, space) for the testing, development and implementation of the migration between components
Supporting users	User support	Individual knowledgeable in data and system
	User statistics	Maintain information on who requests data, number of requests and volume requested
	Corrections and additions	Contact past requestors concerning errors and updates
	Independent data QA	Conduct formal data review
	Value-added products	Optional: create integrated sets of data
	Database tools	Optional: provide analysis and graphical display tools
	Advertising	Market data through scientific outlets and on Web
	Long-term funding support	Maintain sponsor contact to ensure continued funding

6.5.3 Ingesting data

The archive staff may assign keywords based on the metadata to be used in search and order functions. This activity also includes defining the directory and filename structure and syntax.

6.5.4 Data preservation

One of the prime responsibilities for a data archive is to make sure that the data will be accessible over a long time—beyond the life of projects and

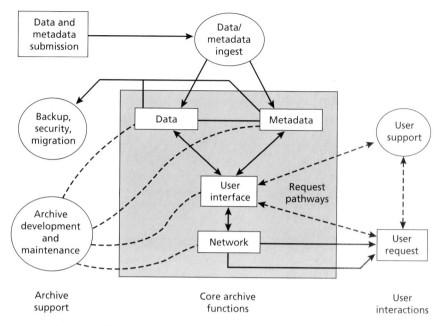

Fig. 6.2 The flow of data and metadata from data submission through user requests showing the internal archive and system functions. Solid lines indicate the primary flow while dashed lines indicate actions that periodically occur as needed or requested.

programmes. The ESA FLED report (Gross *et al.* 1995) identified many examples of lost data. Just because data were put in computerized format does not mean the data archive problem is solved. For example, the International Biological Program (IBP) data were stored on electronic media in the early 1970s with the promise of long-term availability, but they are no longer accessible. A price of continual improvement in computer technology is the obsolescence of the older technology. In fact, the IBP data were recently re-entered by a project at ORNL from faded computer printouts after the 'permanent' magnetic media became unusable. The lessons are that, firstly, maintaining paper copies of valuable data sets is a worthwhile practice, and secondly that planning to continually upgrade electronic storage is required (e.g. transferring from 8-inch disks to 5¼-inch floppy diskettes to CD-ROMs to future media).

Back-up and security

Backup and security are important for insuring the longevity of the archived data and metadata. Even the most carefully managed information systems risk information loss from mistakes, accidents and malicious actions. A backup is a copy of the information. This extra copy should be stored by a device with media that is independent of the primary storage system. Additional protection is provided when the extra copy is stored in a separate location (e.g. another building). The backup logic must be tested with 'test retrievals'. A common operations adage is 'Backup is easy, restoration is another matter'. When archive technology is migrated to new hardware or software, continued access to the backup device or media must also be considered.

Backup (and limited access to the backup copy) contributes to the security of the information. Security also includes procedures to limit and record the transactions that can change the data or metadata. These procedures must be implemented for system, network, web server and database software. Self-surveillance should be routinely conducted to verify the integrity of the archive. The security of the archive must be evaluated based on the 'replacement cost' of the information. Because the past cannot be revisited, most ecological data are irreplaceable. Evaluating the security needs of the archive based on the 'sensitivity' of the information (e.g. locations of collectable endangered species or measurements of hazardous pollutants) may lead to an insufficiently protected system. While 'sensitive' information must be protected, other information can also become the target for random mischief.

Paper archive

Guidelines for creating paper copies for long-term storage have been described by Lehman (LTER 1996; and personal communication, Clarence Lehman, July 1997). The idea is to apply to paper records the same kind of error detection used in other media, like magnetic disks. A paper copy should be readable by humans and scannable by computer, with the accuracy expected from computer media. The Printing using Error Recovery Method 1 (PERM1) (LTER 1996) includes suggestions for using acid-free paper, permanent ink, and printing check sums on each sheet.

Even given the seemingly wonderful properties of CD-ROMs, it is predictable that they will eventually be replaced by another technology. Rothenberg (1995) points out hardware and software considerations to ensure that future generations can successfully read digital files, such as avoiding data compression, avoiding proprietary storage formats and devising bootstrap standards to use as input to future programs. This periodic upgrading of storage media (something that libraries have not had to address with books) requires both long-term planning and funding. The current life cycle is 3–5 years for 'current technology' and ~10 years for available technology. The reliability of extending the functional lifecycle beyond these limits should be carefully considered.

Digital storage

There is no universal way to preserve data and metadata in a digital format

Table 6.7 Examples of file types and standard extensions.

Format type	Format	Extension
ASCII	Vertical bar delimited	.DAT
ASCII	Comma separated values with quoted strings	.CSV
ASCII	Fixed column (system data format)	.SDF
ASCII	Tab delimited	.TAB
Database	Access	.MDB
Database	dBase	.DBF
Database	FoxPro	.DBF
GIS	ARC/INFO Export file	.E00
Spreadsheet	Lotus 1-2-3 version 2	.WK1
Spreadsheet	Excel	.XLS
Spreadsheet	Data Interchange Format (ASCII)	.DIF
Statistical/DB	SAS Cport file	
Structured	Network Common Data Form (NetCDF)	.CDF
Structured	Hierarchical Data Format (HDF)	.HDF

(Table 6.7). The media and format have changed and will continue to change with evolving computer technology. One of the recommended practices is to avoid proprietary packages. Currently, ASCII flat files appear to be the most stable, neutral format for both data and metadata. Most commercial packages have built-in functions to output ASCII files, such as CSV from spreadsheets.

A drawback of ASCII formats is that a tedious, error-prone process is required to read an ASCII file into a software package (spreadsheet, data management system, DBMS) to use the data. It is difficult to detect format errors, such as reversing columns. It is also difficult to detect problems with the completeness of the ASCII data documentation.

Several self-documenting, semi-neutral formats are emerging, including NetCDF and HDF. NetCDF (Network Common Data Form) is an interface for scientific data access that implements a machine-independent, self-describing, extendible file format (URL:http://www.unidata.ucar.edu/packages/netcdf/). NetCDF has been developed by the University Consortium for Atmospheric Research (UCAR) for atmospheric and other data types (Rew *et al.* 1997). It is a widely used, robust binary format with a variety of database functions and subroutines.

Hierarchical Data Format (HDF) was developed by the National Center for Supercomputing Applications (NCSA) to assist users in the transfer and manipulation of scientific data across diverse operating systems and machines. HDF is a self-defining file format that supports a variety of data types: scientific data arrays, tables and text annotations, as well as several types of raster images and their associated colour palettes (http://hdf.ncsa.uiuc.edu/).

6.6 Locating data

Effectively representing the contents of the archive to the potential user community is a significant objective. Typically archive staff are a mix of systems specialists, database managers, information specialists and scientists. It is essential that scientists be involved because they play a critical role in the organization and presentation of data, QA/QC review of data and metadata and development of value-added products. Success of the communications between the archive and user community depends on the proper selection of a documentation structure (how data sets are categorized) and language (what keywords are used). Input from scientists about the suitability of documentation is very important. In addition, a scientific advisory group can provide a valuable link among users, sponsors and staff.

6.6.1 Index system

The computing technology associated with the index or catalogue is typically a database management system. The constructed database contains key fields and words to allow for tabulation of data attributes (e.g. How many data sets or records of what type? From what place? From what time?), and querying of data set records according to user-specific criteria. The database technology required for this function of the archive includes relational databases (and report generation utilities) or more integrated data analysis software such as SAS, SPSS, etc. The database software may include functions that respond to user-originated queries with either the display of selected data records or summary reports. The database may also be required to record attributes of the users (name, e-mail address, telephone number, etc.) and file access processing (FTP logs, mass storage staging status). Secondary analysis tools should also be included in the requirements for database functionality. Statistical and graphical summaries of data stored or data accessed are helpful for describing the contents of the archive, identifying data of interest and justifying continued operations.

6.6.2 User interface

The user interface is the final and most visible component of the computer technology associated with an archive. The interface provides descriptions of the archive contents to potential data users. These descriptions can be organized in hierarchical lists, for example, sample locations within field sites within eco-regions and measurement types within environmental discipline within temporal or spatial scales. In some cases, the structural relationship of the data sets is too complex to be represented as a hierarchy and the descriptions of archive contents are presented after a database search on keywords, time intervals or geographic regions. The user interface must be able to provide the following major functions:

1 The implementation of links between different levels of a hierarchy (HTML protocols support this processing in Web technology).

2 Entry and display of database parameters. The entry function provides a mechanism for the interface to record user preferences for query criteria. The display function processes the output of query results. This output includes simple statistics (number of matching records), record display and bulk text (sections of documentation may be encompassed in the database).

3 Display of multimedia information, including maps, graphs, text, audio, animation and video. The documentation of ecological data frequently includes non-text information types.

4 Transmit information via the Internet. This information includes the text and graphics associated with data documentation (i.e. by way of Web technology). Information exchange also includes the transfer of data files by way of FTP.

The user interface is typically implemented with locally developed software that incorporates logic provided by applications software such as database servers, forms generation utilities and image display tools. The Web protocols are a common mechanism for implementing the functions of the user interface.

6.6.3 Subsetting data

Archives that store information in databases may include logical functions that allow users to fully specify the desired data subset (i.e. functions that 'clip out' the exact records needed by the user). Archives implemented as collections of files or data sets (groups of loosely related files) are unlikely to generate subsets with consistent and complete structures. The subsetting logic may include the ability to select data by spatial, temporal and thematic criteria. The end-product of subsetting is customized extraction of information for the user (i.e. only those data records matching the time period, locations, parameters measured and measurement range). The subsetting function is a particularly important feature for archives containing very large data objects (e.g. satellite images) or very long sequences of similar information (e.g. weather or stream flow records).

6.6.4 Data delivery

When the user has identified 'data of interest', the archive is responsible for delivering the data. Options for delivery include network, tape and diskette.

Network delivery can be enabled by Web transfer or FTP protocol. Implementations of network transfer must evaluate the quantity of data, the data transfer capacity of the network connection and the reliability of the network connection. If network options are insufficient, then tape or diskette options should be considered. Tape should be selected when the quantity of data is large, access can be deferred (i.e. shipping time) and tape technologies are compatible. The most difficult aspect of the tape option is identifying compatible tape hardware and formats between the archive and the user. Several tape media types are available and each has performance options that may be proprietary for the tape drive hardware or computer operating system. When tape delivery is used, documentation procedures for tape options must be implemented. For small quantities of data, diskette is a delivery option when the network option is not available. Diskette technology includes floppy disks, disk cartridges and CD-ROM. Each of these media has a limited number of format options.

6.7 System development

Computer technologies as a component of data archives are not discussed in detail here; however, the advances in this area, especially PCs, networks and the World Wide Web, contribute greatly to the growth of data archives. The elements discussed in this section: storage system, information system and a user interface, depend directly on computer technology.

The data storage systems are the primary computing technology associated with the data and metadata in an archive. Table 6.8 describes a few of the storage system options.

6.8 Supporting users

Other functions of the archive that may enhance value to the user community include conducting QA on the data and metadata beyond that performed by the original data compiler, and combining archive data with other data. These other data may be an integrated collection of data (a value-added data product) contributed by synthesis and modelling projects. Processing by the archive to promote the consistency and completeness of data (e.g. converting to SI metric units) will enhance the value of the data.

In addition to maintaining a long-term, secure data archive, the archive staff will also provide post-project support, such as answering user questions, identifying sources of reference information, informing users of updates and additions, and maintaining user statistics. Archives can promote the growth and value of their data holdings through a strategy for incorporating data updates, value-added products (especially from synthesis and modelling applications), and user feedback; and by portraying the integrating relationships between

Table 6.8 Computing technology options for storage systems (GB = Gigabyte or approximately a trillion bytes).

Attribute	Small	Medium	Large
Size	A few GB	10s–100s GB	1000s GB
Primary hardware	A single disk	Many disks organized into a RAID unit	Mass storage system with RAID units, automated tape robots, tape storage libraries, multiple host computers for interface control
Primary software	Computer operating system	Computer operating system and RAID controller software	Computer operating system, RAID controller software and Mass storage system software
Accessibility	Online	Online	Staged online and near-line (accessible after tape request and tape mount)
Information protection	Requires backup scheme	Various levels of redundancy can be configured within RAID logic, backup may be required	Duplicate tape copies possible, master directory of files requires backup / recover protection
Technology update	Transfer to new device is simple	Transfer to new device is simple, but can require several hours-days to complete	Transfer to new device is complex and may require several days-months to complete
Support	Competent computer user	Part-time system manager	One or more full-time system managers
Hardware cost	$100s–1000s	$10 000s–100 000s	$100 000s

new and existing data. Staff may collaborate with scientists to determine useful enhancements to data sets (e.g. add common variables, aggregate to common units, or calculate uncertainty based on user needs). In addition, understanding the scientific intent of current and future use of the data enables the staff to optimize the maintenance of the data and the design of the archive. It is also crucial for the staff to promote the availability of the data by interacting with the user community during attendance at professional meetings and utilizing marketing techniques.

6.9 Future directions for data archival

The use of ecological data will continue to expand through future synthesis and assessment studies addressing long-term regional and ecological issues. Many future ecological discoveries will depend on the intense analysis of very

> **Box 6.1 Suggested actions for increasing data archiving and sharing**
>
> - Provide incentives for sharing and archiving data
> - Recognize data sets with metadata as valuable research products
> - Establish a universal citation policy for data
> - Establish guidelines or expectations for metadata
> - Develop data distribution and archive centers
> - Ensure long-term financial and institutional support for data archives
> - Develop and widely adopt discipline-specific guidelines for metadata and codes (e.g. like CAS numbers).

large aggregations of data. Further reductions in cost per unit of electronic storage will inspire opportunities for expanded archives (storage capacity per unit hardware cost has decreased 1000 times in the past 10 years and this trend is expected to continue). The processes and practices employed for ecological data archiving must also expand to prevent the establishment of 'bigger data chaos' and to facilitate new ecological discoveries. In addition to large data aggregations, more value-added products, such as integrated data sets for modelling, will become available. More tools will be available for documenting data and they will be used routinely. Sharing and archiving data will be more efficient if the following general principles are considered in overall project planning and operations:

1 establish the flow of data from investigator to a formal data archive as part of the work plan;

2 process data and metadata to achieve optimal consistency and completeness; and

3 institute policies to give adequate credit to the data producers for their archival efforts.

Specific actions that can facilitate data archiving and sharing are listed in Box 6.1.

6.10 References

Esser, G., Lieth, H.F.H., Scurlock, J.M.O. & Olson, R.J. (1997) *Worldwide Estimates and Bibliography of Net Primary Productivity from Pre-1982 Publications.* ORNL/TM-13485. Oak Ridge National Laboratory, Oak Ridge, TN.

Gross, K.L., Pake, C.E. & the FLED Committee Members. (1995) *Final Report of the Ecological Society of America Committee on the Future of Long-term Ecological Data (FLED), Vol. I. Text of the Report.* Ecological Society of America. Washington, DC.

Keeling, C.D., Bacastow, R.B. & Whorf, T.P. (1982) Measurements of the concentration of carbon dioxide at Mauna Loa Observatory, Hawaii. In: *Carbon Dioxide Review.* (ed. W.C. Clark), pp. 377–385. Oxford University Press, New York.

Long-Term Ecological Research Network. (1996) *Data Management Committee Report of Annual Meeting*. LTER Network Office, Albuquerque, NM. (available from http://www.lternet.edu/documents/Reports/Data-management-committee/1996/)

Michener, W.K., Brunt, J.W., Helly, J., Kirchner, T.B. & Stafford, S.G. (1997) Non-geospatial metadata for the ecological sciences. *Ecological Applications* **7**, 330–342.

Moore, R.V. (1997) The logical and physical design of the Land Ocean Interaction Study database. *The Science of the Total Environment* **194/195**, 137–146.

National Research Council. (1991) *Solving the Global Change Puzzle, A US Strategy for Managing Data and Information*. Report by the Committee on Geophysical Data Commission on Biosciences, Environment, and Resources. National Academy Press, Washington, DC.

National Research Council. (1995a) *On the Full and Open Exchange of Scientific Data*. National Academy Press, Washington, DC.

National Research Council. (1995b) *Finding the Forest in the Trees: The Challenge of Combining Diverse Environmental Data*. National Academy Press, Washington, DC.

National Research Council. (1995c) *Preserving Scientific Data on Our Physical Universe: A New Strategy for Archiving the Nation's Scientific Information Resources*. National Academy Press, Washington, DC.

National Research Council. (1997) *Bits of Power: Issues in Global Access to Scientific Data*. National Academy Press, Washington, DC.

O'Neill, H.J., Bingham, R.A., Howell, G.D., Duerden, F.C. & Roberts, C. (1993) Environmental auditing: development of an integrated environmental database. *Environmental Management* **17**, 257–265.

Peet, R.K. (1998) ESA Journals: evolution and revolution. *Bulletin of the Ecological Society of America* **79**, 177–181.

Porter, J.H. & Callahan, J.T. (1994) Circumventing a dilemma: historical approaches to data sharing in ecological research. In: *Environmental Information Management and Analysis: Ecosystem to Global Scales*. (eds W.K. Michener, J.W. Brunt & S.G. Stafford), pp. 193–202. Taylor & Francis, Ltd., London.

Rew, R.K., Davis, G.P., Emerson, S. & Davies, H. (1997) An Interface for Data Access, Version 3, April 1997. *NetCDF User's Guide for C*. (Available from UCAR Unidata Program Center P.O. Box 3000 Boulder, Colorado, USA 80307 or in PostScript form by anonymous FTP from URL:ftp://ftp.unidata.ucar.edu/pub/netcdf/guidec.ps.Z).

Rothenberg, J. (1995) Ensuring the longevity of digital documents. *Scientific American* **272**, 42–47.

Sauer, J.R., Hines, J.E., Gough, G., Thomas, I. & Peterjohn, B.G. (1997) *The North American Breeding Bird Survey Results and Analysis*, Version 96.4. Patuxent Wildlife Research Center, Laurel, MA.

Slagel, R.L. (1994) Standards for integration of multisource and cross-media environmental data. In: *Environmental Information Management and Analysis: Ecosystem to Global Scales*. (eds W.K. Michener, J.W. Brunt & S.G. Stafford), pp. 222–223. Taylor & Francis, Ltd., London.

Strebel, D.E., Meeson, B.W. & Nelson, A.K. (1994) Scientific information systems: a conceptual framework. In: *Environmental Information Management and Analysis: Ecosystem to Global Scales*. (eds W.K. Michener, J.W. Brunt & S.G. Stafford), pp. 59–85. Taylor & Francis, Ltd., London.

CHAPTER 7

Transforming Data into Information and Knowledge

WILLIAM K. MICHENER

7.1 Introduction

Designing experiments that incorporate adequate replication, controls and interspersion of uniform treatments can be a relatively straightforward process in many scientific disciplines. Ecologists, however, must frequently work in situations where circumstances are less than ideal. For instance, experimental units are often heterogeneous and true 'replication' may be difficult or impossible. 'Treatments' may be non-standard and consist of multiple treatments applied iteratively (e.g. ecological restoration projects), as opposed to a single treatment. Natural and anthropogenic disturbances as well as other confounding factors, the consequences of which are poorly understood, may be common and result in many data idiosyncracies. Furthermore, the variability in system responses may be of greater interest and ecological significance than mean responses. In some cases, for example, we may be interested in documenting the spatial extent and variability associated with an ecological response, rather than precisely quantifying the response at a single point. Additional issues related to the design of ecological research projects are discussed more fully in Chapter 1 and the references therein.

As a consequence of environmental stochasticity and other factors previously alluded to, ecologists must be especially creative and diligent in their efforts to isolate drops of knowledge and streams of information from an ever-expanding ocean of data. The difficulties inherent in truly repeating ecological experiments are such that ecologists frequently rely on several lines of evidence. For instance, data often are synthesized from multiple ecological experiments in the search for general patterns. Because experiments often differ in one or more ways, ecological experiments may need to be supplemented with observations from long-term studies, simulation models or other approaches described in Chapter 1.

An enormous number of sophisticated and diverse processing and analytical tools are included in the ecologist's toolbox—far more than could ever be comprehensively discussed in a textbook, much less a single chapter. However, a treatment of ecological data would be incomplete without at least brief consideration of how data are transformed into the information and knowledge that were, presumably, the objectives of the study in the first place.

Consequently, some of the steps involved in managing and processing data are reviewed in the following sections. A small and, admittedly, subjectively selected subset of graphical and statistical analytical approaches available to ecologists are briefly described along with pointers to relevant literature. Finally, several relatively new analytical approaches that are suitable for very large databases are introduced.

7.2 Data management and processing

Data and information management is a process that starts with project design and can extend well beyond the data analysis and publication phases. Information management within an organizational context can include aspects of project design and planning (Chapter 1), design of paper and digital data entry forms (Chapter 2), quality assurance and quality control (Chapters 2 & 4), data processing (e.g., subsetting, merging; Chapters 2 & 3), metadata development (Chapter 5), and submission of data and metadata to a data centre or data archive (Chapter 6). Generally, a successful long-standing research programme at an institution depends on a high-quality data management programme that comprises all of the components described above (e.g. Briggs & Su 1994; Ingersoll *et al.* 1997; Strebel *et al.* 1994).

Data, as described in Chapter 1, consist of a series of rows and columns of alphanumeric characters (Fig. 7.1a). Consequently, data have little or no intrinsic information content and must undergo processing (often considerable) to be transformed into information. At a minimum, raw (unprocessed) data usually must undergo some basic processing and must be merged or viewed in conjunction with at least a subset of their associated metadata to derive information (Fig. 7.1b).

Data processing can occur at various scales. Subsetting, merging and other operations discussed in Chapters 2 and 4 can be used to manipulate one or more entire data sets. Data that comprise one or more observations or parameters can also be processed in various ways to meet specific objectives. For instance, data processing can include procedures designed to reduce the volume of data, as well as procedures that transform data values prior to interpretation and analysis.

7.2.1 Data reduction

Data reduction is often necessary when automated data collection procedures are employed. For example, temperature, precipitation and other meteorological sensors can be programmed to collect data at very frequent intervals (seconds to minutes). Such high-resolution temporal sampling can, however, generate extremely large data sets that exceed most scientific requirements.

(a)

0711980500276000
0711980600276000
0711980700277003
0711980800282017
0711980900285000
0711981000293000
0711981100301000
0711981200304000

(b)

Date	Time	Air temp C	Precipitation mm
11July1998	0500	27.6	000
11July1998	0600	27.6	000
11July1998	0700	27.7	003
11July1998	0800	28.2	017
11July1998	0900	28.5	000
11July1998	1000	29.3	000
11July1998	1100	30.1	000
11July1998	1200	30.4	000

Fig. 7.1 (a) Unprocessed meteorological data set. (b) Incorporation of a subset of metadata and minimal processing (i.e. temperature/10) provide structure and some meaning to the data, although the context (location, research design, etc.) and other metadata attributes are absent.

Consequently, high-resolution temporal data are often filtered to a coarser scale. For example, data collected at intervals on the scale of minutes or seconds may be reduced to hourly means, minima and maxima. In some cases, the data reduction procedure takes place in the field using programmable data loggers to optimize remote data storage (i.e. minimize number of trips to the field to download data). In other cases, data reduction can occur in the laboratory and both the reduced data set and the complete version may be available for further analysis. The degree to which data are reduced in volume is affected by data storage capabilities and costs, computational time required for data analysis, and other factors that must be weighed on a case-by-case basis (see also Chapter 2).

7.2.2 Data transformation

Data must frequently be transformed from one format to another in order to be more readily interpretable. Simple examples include single-step conversion of measurement units (e.g. metric conversion), scale and other parameters (see Schneider 1994). Somewhat more complicated multiple-step data transformations may also be necessary. For example, plankton counts are often based on multiple sub-samples from one or more samples that were collected by one or more nets pulled through the water for various distances or times. The raw counts must be transformed to a standard measure (e.g. number of organisms per cubic metre of water sampled) to be comparable. Such transformations may require the incorporation of one or more conversion factors (e.g. flow metre calibration), each of which should, ideally, be thoroughly documented. By maintaining a record (metadata) of the various conversion factors and algorithms employed, it is possible to back-calculate to the raw data from the derived values.

Many data transformations are simple linear conversions such as multiplication by a constant value. At the other extreme, transformations can involve multiple steps and a mixture of linear and nonlinear algorithms. A common example is the digital image processing steps necessary to convert 'raw' spectral data from satellite sensors to vegetation indices like the Normalized Difference Vegetation Index that are geometrically and atmospherically corrected (Jensen 1996). Generally, the more complicated the transformation, the more important it is to maintain a detailed history of the steps performed. In addition, it is often useful to archive one or more intermediate data sets, particularly for complex and computer-intensive (expensive) transformations.

7.3 Graphical and statistical analyses

The choices among space, time and thematic parameters that are made in designing a research project (see Chapter 1) also affect the choice of statistical analyses. For instance, conventional statistics (e.g. ANOVA, linear regression) may be entirely appropriate for studies having a modest number of sites and parameters and a low sampling frequency, whereas geostatistical techniques may be desired for analysing data collected at a large number of sites. Conventional graphical and statistical analyses, as well as several general classes of statistical approaches more specifically suited to temporal, spatial or thematic sampling, are briefly described in the following sections. In addition, several other relatively new or under-utilized statistical approaches that may be relevant to particular ecological analyses are presented (see also Michener 1997 and Piegorsch *et al.* 1998).

7.3.1 Graphical analyses

Graphics are used for several different purposes in ecological research. First, graphics can play an important role in data quality assurance by helping to identify unusual values and potential data contamination (see Chapter 4). Second, graphics can be extremely valuable in allowing scientists to visually explore their data. The patterns are often obscured in data tables and statistics

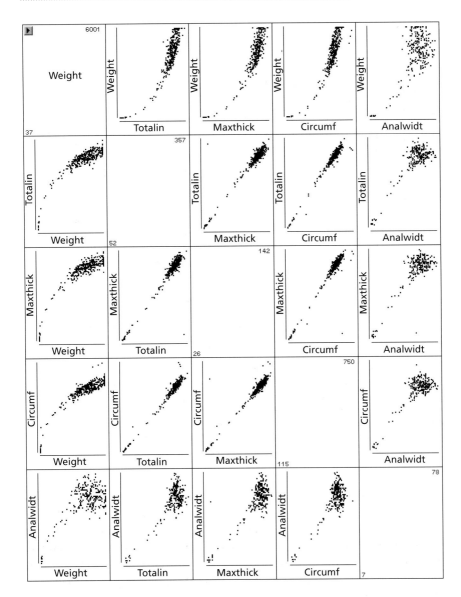

Fig. 7.2 X–Y scatterplots of gopher tortoise (*Gopherus polyphemus*) morphometrics. (Michener, unpublished data.)

but can be readily apparent in a few well-chosen graphical representations. The use of graphics in searching for patterns in data is referred to as exploratory data analysis. Some especially useful graphical approaches include X–Y scatter plots (where all variables are plotted against one another; Fig. 7.2), and box-and-whisker plots, which depict the statistical distribution of the data (Fig. 7.3). In addition, some recent statistical and graphical software products

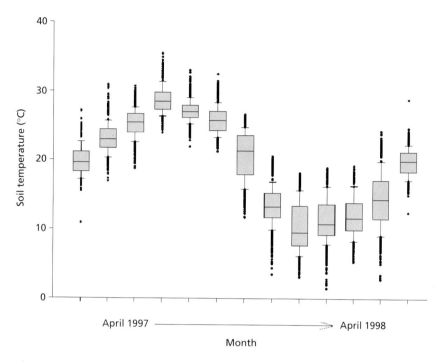

Fig. 7.3 Box plots depicting the statistical distribution of soil temperature data (Simkin, Michener, & Wyatt, unpublished data). Upper (75%) and lower (25%) quartiles (Q) are portrayed by the top and bottom of the rectangle; median is portrayed by horizontal line within the rectangle, lines outside boxes extend to the upper adjacent value (the largest observation ≤ upper quartile (Q(75%)) plus 1.5 × interquartile range (IQR; = Q(75%) − Q(25%))) and lower adjacent value (the smallest observation ≥ Q(25%) − (1.5 × IQR)); values falling outside the range of adjacent values are plotted as individual points.

allow interactive exploration of a data set. For instance, in Fig. 7.4 are a series of multiple screens produced by SAS Insight®: a scatter plot (temperature data versus time) with one specific observation highlighted, a listing of the values for that observation and a portion of the worksheet composing the underlying data table. Excellent references for exploratory data analysis techniques include Tukey (1977), Chambers *et al.* (1983), Hoaglin *et al.* (1983), Cleveland (1985), and Ellison (1993).

Graphical approaches for quality assurance and exploratory data analysis such as those shown in Figures 7.2–7.4 (also see Chapter 4) are underutilized by ecologists (Ellison 1993). The most common use of graphics by ecologists is for conveying information in presentations and publications. Maps of study sites, X–Y graphs and bar charts are now commonplace in ecological publications. Tufte (1983, 1990), Tufte and Krasny (1997) and Ellison (1993) provide useful guidelines and examples of high-quality graphics that convey a maximum amount of information with minimal clutter. In addition to conventional

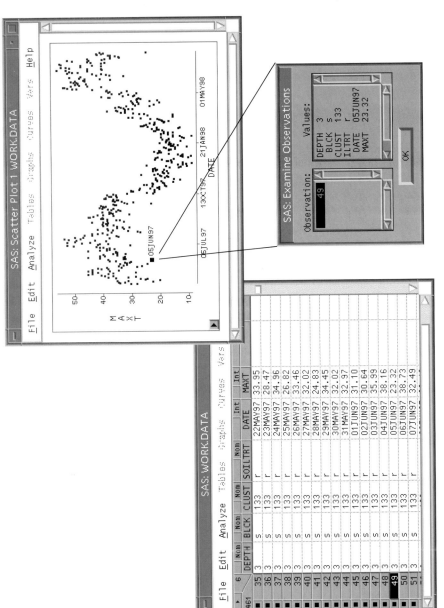

Fig. 7.4 A series of multiple screens produced by SAS Insight® that show (a) a scatterplot (maximum soil temperature data vs. time) with one specific observation highlighted; (b) a box that lists the values for that observation; and (c) a portion of the worksheet that comprises the underlying data table. (Simkin, Michener, & Wyatt, unpublished data.)

graphics produced by most statistical and graphical software products, advanced Geographic Information System software (e.g. ARC/Info®) and scientific visualization programs (e.g. IBM Data Explorer®) allow one to produce dynamic, multidimensional, and animated graphics. Helly (1998) discusses the production steps involved in scientific visualization and presents several relevant examples from ecology. For general references to scientific visualization see Brown *et al.* (1995), Gallagher (1995) and Nielson *et al.* (1997).

7.3.2 Conventional statistics

Ecologists generally receive some formal training in statistical techniques that can be applied in experimental studies. Such techniques include, but are not limited to, analysis of variance (ANOVA), multivariate analysis of variance (MANOVA), simple linear and multiple regression and analysis of frequencies (e.g. Goldberg & Scheiner 1993; Potvin 1993; Scheiner 1993; Shaw & Mitchell-Olds 1993; Sokal & Rohlf 1995). Common to most of the methods are the following assumptions: sampling is random; error terms are random normal variables; error terms are independently distributed (i.e. not correlated in space or time); and variances are homogeneous. The statistical methods referenced above are primarily applied in situations where experimental manipulations or treatments are applied to replicated study units and a priori hypotheses are tested. In addition to the literature cited above, several textbooks provide excellent sources for additional information (e.g. Draper & Smith 1981; Myers 1986; Winer *et al.* 1991; Neter *et al.* 1996; Zar 1996).

7.3.3 Descriptive statistics

Much ecological research is focused on describing and elucidating patterns observed in nature. Frequently, experimental manipulations are not or cannot be performed, and a priori hypotheses may or may not be tested. It may, in fact, be difficult or inappropriate to ascribe probability concepts to many descriptive statistics. Typical patterns of interest to ecologists include the distribution of organisms in space; species-abundance relationships; interspecific associations; community structure, similarity and dissimilarity; and community–habitat relationships. Among the statistical methods that might be categorized as observational or descriptive are diversity indices, cluster analysis, quadrat variance and distance methods (for characterizing patterns), principal components analysis, correspondence analysis and various ordination techniques (Gauch 1982; Digby & Kempton 1987; Jongman *et al.* 1987; Ludwig & Reynolds 1988; Buckland *et al.* 1993; Manly 1994).

Many of the techniques used for generating descriptive statistics are appropriate for studies based on a small number of observations of many parameters (e.g. all species in a community at a site). For example, various ordination techniques are used to summarize the relationships between different species based on their abundance at different locations or, conversely, are used to summarize the relationships between different locations based on the abundance of different species at those locations (Manly 1994). Ordination can be accomplished using principal components analysis, principal coordinates analysis, multidimensional scaling (both metric and nonmetric) and different types of correspondence analysis (Jongman *et al.* 1987; Manly 1994). In addition to comprehensive statistical packages that often include one or more descriptive approaches, specialized software programs are frequently used by ecologists for canonical correspondence analysis (e.g. CANOCO, ter Braak 1987), detrended correspondence analysis (e.g. DECORANA, Hill 1979a; Hill & Gauch 1980), cluster analysis (TWINSPAN, Hill 1979b) and related analyses.

Cost, personnel and other constraints frequently limit the number of sites and points in time where sampling can be 'replicated'. Although these descriptive statistics may be appropriate for characterizing patterns, focused experimentation and other research approaches may be necessary for identifying causal mechanisms.

7.3.4 Temporal analytical approaches

Many questions related to ecological structure and function can only be addressed through repeated sampling of salient parameters at multiple points in time. Depending on the question being 'asked', an ecologist may sample an experimental unit after a single treatment to characterize subsequent change; sample multiple experimental units after 'similar' treatment(s) at various times in the past (i.e. space-for-time-substitution); sample an experimental unit after repeated treatments; sample an experimental unit pre- and post-treatment; or sample an experimental unit and control(s) pre- and post-treatment (i.e. Before-After-Control-Impact (BACI) study; Stewart-Oaten *et al.* 1986). Various time series analyses as well as conventional, observational and other statistical techniques discussed in the remainder of this section may be appropriate statistical methods. However, it should be noted that some approaches are more relevant to characterizing temporal patterns, whereas others support explicit hypothesis testing.

Many traditional time-series analyses are designed to decompose temporal series of observations into overall trends (e.g. mean), seasonal cycles and irregular fluctuations (Chatfield 1984). Some of the most basic statistical models were initially developed to support analyses of temporal records where the overall mean and variance are not stationary. Time series are characterized as stationary if there are no systematic changes in the mean and variance and if there are no periodic variations (e.g. seasonal variation). For example, moving average (MA) models are appropriate for situations where a stochastic event may have both an immediate effect, as well as effects in several subsequent time periods; that is, observations at a particular time are related to current and past random errors. Autoregressive (AR) models are designed for situations where response variables measured at a particular time are correlated with previously measured values. A third group of models, autoregressive moving average (ARMA) can be applied to stationary time series affected by both autoregressive and moving average processes.

Most ecological time series are nonstationary (e.g. Fig. 7.5a). In such cases, autoregressive integrated moving average (ARIMA) models are employed to remove or quantify non-stationary sources of variation. Time series consisting of at least 50 observations are frequently necessary for ARIMA models, and extremely long time series may be required to elucidate complex processes operating at seasonal, annual and decadal scales. Spectral analysis, contingency periodograms, and principal components analysis may also be used to characterize cyclical variability in time series (Chatfield 1984; Jassby & Powell 1990; Cloern & Jassby 1995; Stockwell *et al.* 1995). These three techniques are primarily used to identify or quantify the magnitude of periodicity (e.g. diurnal, seasonal, annual, decadal) in a time series. These and related analyses may be especially useful for determining the length of a time-series record needed to characterize temporal variability prior to an experiment.

Several semiparametric approaches, often referred to as 'robust regression' techniques, may be applicable to time series where the relationship between the dependent variables and the independent variable (time) does not follow a single functional form (e.g. Cleveland & Devlin 1988; Wahba 1990). These techniques employ a locally weighted smoothing function that varies over the time series. Robust time series regression techniques have been applied to quality assurance/quality control (QA/QC) of ecological data (Chapal & Edwards 1994) and Trexler and Travis (1993) discuss locally weighted regression techniques that are applicable to non-time series data.

In addition to cyclical and random processes, ecologists are often interested in non-random changes that accompany 'treatments'. In such cases, change may be manifested as a simple discrete step function or as a more complex nonlinear response (e.g. exponential decay). The ecologist may, for example, be interested in determining whether or not the rate of a functional process has changed in response to a particular treatment (e.g. Fig. 7.5b). Intervention analysis is a statistical technique, based on ARIMA, that allows the scientist to specify when an intervention or treatment occurred, and then compare post- versus pre-treatment trends (Carpenter 1990; Rasmussen *et al.* 1993). For example, time-series analysis would be used to characterize trends and seasonality or other periodicity in a long-term record such as a hydrological time series (Fig. 7.5a); however, intervention analyses would quantify changes in

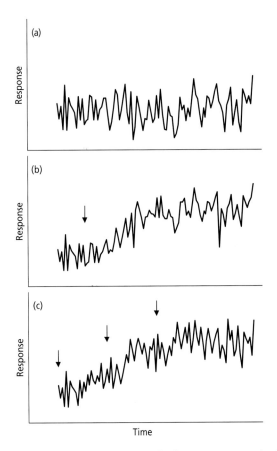

Fig. 7.5 (a) Example of a non-stationary, seasonally fluctuating time series for a parameter observed at a single location; (b) Example of a nonstationary, seasonally fluctuating time series for a parameter observed at a single location that is subjected to a single treatment (intervention; at arrow). (c) Example of a nonstationary, seasonally fluctuating time series for a parameter observed at a single location that is subjected to three repeated treatments. (From Michener 1997, by permission of Blackwell Science.)

trends or periodicity in response to imposition of a disturbance or treatment such as the introduction of a water control structure (Fig. 7.5b). Intervention analysis offers both advantages and disadvantages. Like the general classes of time series models previously discussed (AR, MA, ARMA, ARIMA), intervention analysis does not require that experimental units be replicated. Furthermore, the ability to compare pre- and post-treatment temporal trends has obvious relevance to landscape and restoration ecology where relatively large treatments may be applied at one or a few discrete points in time. Several disadvantages are also common to the various models. First, the time series of observation prior to and following the treatment need to be sufficiently long to account for seasonal and other cyclical trends. Second, it is important to remember, especially for unreplicated time series, that correlation does not imply causation (i.e. responses to a treatment may be entirely coincidental). Several approaches may be employed to increase the likelihood that changes observed in response to a treatment are not due to chance. These include observations of consistent responses to multiple treatments (e.g. fire, fertilization, thinning) over time at a single site (e.g. Fig. 7.5c) and comparison of pre- and post-intervention trends with similar data from one or more controls or reference sites (if available).

Ecologists may not have the option of choosing a 'treatment' site(s) or establishing long-term sampling programmes prior to the imposition of 'treatments'. In such cases, many of the time-series approaches presented previously are inappropriate or invalid. Several analytical designs have been developed for dealing with these situations. Many, however, do require that some data be collected prior to the 'treatment'. For example, Before-After-Control-Impact (BACI) studies were designed to compare a single treatment site with a single control site, using data collected prior to and following the treatment (Fig. 7.6a). In BACI studies, it is possible to determine whether or not the mean values of the response variable(s) changed in response to the treatment (Green 1979; Stewart-Oaten *et al.* 1986). Optimal BACI design strategies are discussed more fully by Stewart-Oaten *et al.* (1986, 1992).

BACI designs have been criticized for their inability to discriminate treatment effects on variance and their lack of spatial replication. Underwood (1991, 1992, 1994) suggests these problems can be minimized or eliminated by using asymmetrical designs ('beyond BACI') where a treatment site is compared to multiple control sites (Fig. 7.6b). Where pre-treatment sampling is impossible, Wiens and Parker (1995) discuss several *post facto* designs that compare impact level-by-time, impact trend-by-time, as well as gradient or trend studies, using several relevant ecological examples from the *Exxon Valdez* oil spill. For example, impact level-by-time interaction analysis can be used to compare profiles of means for treated and reference sites over time and may include pre-treatment data if available. Figure 7.6c illustrates an impact level-by-time design where repeated measurements are made over time at a set of reference sites and compared with two sets of sites that received different levels of treatment; the dashed lines indicating the potential for incorporation of pre-treatment data for before-after comparisons. The level-by-time interaction analysis accounts for differences between treatment and reference areas, but assumes that temporal changes in the response variable are similar between treatment and reference areas in the absence of treatment (Wiens & Parker 1995). Impact trend-by-time designs are applicable in situations where treatments can be measured as a continuous variable and one is interested in comparing trends between measures of the response variable and a spectrum of treatment levels over time.

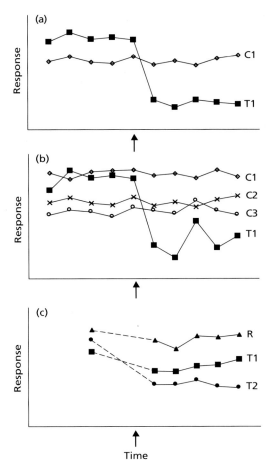

Fig. 7.6 (a) Simulated response (decreased mean, variance unchanged) to a treatment (arrow) applied at one treatment (or experimental) site (T1) in comparison to one control site (C1), each sampled 5X before and after the treatment. (b) Simulated response (decreased mean, increased variance) to a treatment applied at one treatment site (T1) in comparison to three control sites (C1, C2, C3), each sampled 5X before and after the treatment. (c) Simulated responses observed at two treatment sites (T1, T2), each exposed to a different level of treatment, in comparison to a reference site (R). (From Michener 1997, by permission of Blackwell Science.)

7.3.5 Spatial analytical approaches

Most conventional statistical procedures (e.g. regression, multiple regression, ANOVA) assume observations are made under identical conditions and that they are independent (i.e. random sampling). Hypothesis tests, confidence intervals, prediction intervals and other formal inferences may be invalid when the assumption of independence is violated. In environmental data analysis, observations are rarely independent in time or space. Spatial statistics (or geostatistics) are specifically designed for situations where observations are spatially autocorrelated, meaning that the correlation between any two observations depends on the distance between the sampling locations. Kriging represents one of the most common approaches used for spatial analysis (Cressie 1991). The first step in kriging is to fit a semivariogram model to a spatially explicit set of observations to determine the rate at which correlation decreases with distance. Second, semivariogram model results are used to predict values of a parameter over a grid of sites at which it was not observed. Finally, standard contouring algorithms can then be used to graphically depict kriged values. Fifty to 100 or more spatial observations may be required for kriging, depending upon the intrinsic spatial variability of the data. In addition to predicted values, standard errors for the predicted values can also be derived. Thus, it may be possible in some cases to perform a pilot study, identify 'hot spots' of variability (e.g. ecotones) and design a final sampling program where samples are more densely clustered in the areas with the highest variability. Additional details related to geostatistics and kriging are provided in Cressie (1991) and Legendre (1993). Related approaches based on spatial autocorrelation analysis include the GLS-variogram method (Ver Hoef & Cressie 1993) and Mantel and partial Mantel tests (Fortin & Gurevitch 1993). In addition, several excellent GIS texts provide information related to a variety of spatial analytic approaches (Burrough 1986; Burrough & McDonnell 1998; Johnston 1998).

7.3.6 Nonparametric statistical approaches

Nonparametric methods are frequently used in situations where the assumptions of parametric statistical models are violated or there is some doubt about the underlying distribution of the data (e.g. data may not be normally distributed). Common uses of nonparametric methods are for count and rank data, including correlation of paired observations consisting of ranks. However, many statistical analyses (e.g. regression) can be approached using either parametric or nonparametric methods. Comprehensive references to nonparametric approaches include recent texts by Conover (1998), Hollander and Wolfe (1998), and Sprent (1993).

7.3.7 Other relevant statistical approaches

Several additional statistical approaches, many of which are relatively new and others relatively underutilized by many ecologists, may be required for dealing with special cases in ecology. For example, logistic or multiple logistic (for multiple independent variables) regression is appropriate for situations where the dependent variable is categorical, dichotomous, or polychotomous (Hosmer & Lemeshow 1989; Trexler & Travis 1993). Seed germination and plant survivorship studies represent two examples where logistic regression approaches might be appropriate for modelling dichotomous responses. Generalized linear models (GLiMs) represent a special class of models that are appropriate for non-normal count data (e.g. rare species; McCullagh & Nelder 1989; Dobson 1990), and often represent a mixture of parametric and non-parametric methods. Meta-analysis is used to compare and combine outcomes of experiments performed by different investigators (e.g. Gurevitch *et al.* 1992; Gurevitch & Hedges 1993). Other potentially relevant models include repeated measures models based on MANOVA (von Ende 1993); repeated measures models with structured covariance matrices (ARIMA-hybrid) or 'mixed models' (von Ende 1993); nonlinear regression (Box & Draper 1987); and more complex mathematical–statistical models that can accommodate space-time interactions (Haas 1995; Høst *et al.* 1995). Refer to the comprehensive review by Piegorsch *et al.* (1998) for additional recent statistical advances that are particularly relevant for ecologists, including extreme event analysis, adaptive sampling for pollution 'hot spots' and ecological risk assessment.

7.4 Analytical approaches for very large databases

A variety of new tools have recently emerged for processing and analysing data that are contained in very large and/or multidimensional databases. These tools can broadly be classified under two categories:
1 data mining and knowledge discovery; and
2 online analytical processing and data warehousing.

The categories encompass two of the fastest growing segments of the technology market. Although much of the related software has a distinct business orientation, the scientific market is increasingly being addressed. Because of the rapidly emerging technologies, interested readers are encouraged to visit general internet-accessible resource sites (e.g. http://www.kdnuggets.com) and peruse relevant online journals (e.g. *Data Mining and Knowledge Discovery*) for the latest information.

7.4.1 Data mining and knowledge discovery in databases

Business and scientific databases are increasingly becoming so large and messy that direct human analysis and pattern recognition (i.e. knowledge discovery) are precluded. Consequently, computers are being 'trained' to accomplish many analytical tasks that heretofore were largely performed by humans (Mitchell 1997). Knowledge discovery in databases (KDD) has been defined as 'the nontrivial extraction of implicit, previously unknown, and potentially

useful information from data' (Frawley *et al.* 1992). KDD is an iterative process that involves up to six distinct stages, one of which is data mining:
1. develop an understanding of the proposed application;
2. create a target data set;
3. remove or correct corrupted data;
4. apply data-reduction algorithms;
5. apply a data-mining algorithm; and
6. interpret the mined patterns' (Brodley *et al.* 1999).

Relevant examples of KDD include finding specific objects in image data, understanding behavioural preference data, predicting landcover susceptibility to land use change and identifying global change patterns in large, complex databases.

Data mining is associated with grouping similar data, pattern search and development of rules. Various methods including decision trees, neural nets, clustering (or database segmentation), market-basket analysis and deviation detection are employed to mine data (Brodley *et al.* 1999). In all cases, however, domain experts are essential in the mining process and in interpreting resulting patterns. Scientific visualization and related tools are frequently used to facilitate pattern interpretation (e.g. Brown *et al.* 1995; Gallagher 1995; and Nielson *et al.* 1997). More information on KDD and data mining can be found in several recent articles and texts (e.g. Brodley *et al.* 1999; Fayyad *et al.* 1996 a,b).

7.4.2 Online analytical processing and data warehousing

A data warehouse is a central, consolidated database containing all (or most) data from an organization. Within a data warehouse, the data are organized following a multidimensional storage model that can support rapid combinations of data, as well as many different views of the data. A very simple data warehouse might consist of a data cube that is populated by location (X-axis), precipitation (Y-axis), and time (Z-axis). Online analytical processing (OLAP) software allows one to manipulate the data warehouse (i.e. data cube) or other databases, using operations such as slice, dice, consolidate, drill-down and pivot. OLAP can be thought of as turning and twisting a Rubik's Cube of data in different ways to examine various 'what if' scenarios. Whereas conventional transaction processing in relational database management systems deals with individual record-at-a-time processing, OLAP applications usually deal with summarized data that are organized by subject. Furthermore, OLAP applications generally deal with examining trends, thereby requiring the ability to handle time series data. OLAP results can be presented in both tabular and graphical format.

OLAP and data warehousing represent relatively new technologies. High costs for software, in addition to the costs and personnel effort associated with transferring data from existing relational databases into new multidimensional structures in data warehouses have resulted in predominant usage and development for the business world. However, it is likely that the distinctions among OLAP, data warehousing, data mining and knowledge discovery, scientific visualization and related fields will blur, leading to multi-use products that can meet business and scientific needs as well as an array of products that satisfy very specific scientific objectives. For more information on OLAP and data warehousing, see Inmon (1992); Debevoise and Debevoise (1999); Thierauf (1997); and Thomsen (1997). Related software products are described by Flohr (1997) and Thomsen (1997).

7.5 References

Box, G.E.P. & Draper, N.R. (1987) *Empirical Model-building and Response Surfaces*. John Wiley & Sons, Inc., New York.

Briggs, J. & Su, H. (1994) Development and refinement of the Konza Prairie LTER research information management program. In: *Environmental Information Management and Analysis: Ecosystem to Global Scales*. (eds W.K. Michener, J.W. Brunt & S.G. Stafford), pp. 87–100. Taylor and Francis, Ltd., London.

Brodley, C.E., Lane, T. & Stough, T.M. (1999) Knowledge discovery and data mining. *American Scientist* **87**, 54–61.

Brown, J., Earnshaw, R., Jern, M. & Vince, J. (1995) *Visualization: Using Computer Graphics to Explore Data and Present Information*. John Wiley & Sons, Inc., New York.

Buckland, S.T., Anderson, D.R., Burnham, K.P. & Laake, J.L. (1993) *Distance Sampling: Estimating Abundance of Biological Populations*. Chapman & Hall, London.

Burrough, P.A. (1986) *Principles of Geographic Information Systems for Land Resource Assessment*. Oxford University Press, Oxford, UK.

Burrough, P.A. & McDonnell, R.A. (1998) *Principles of Geographic Information Systems*. Oxford University Press, Oxford, UK.

Carpenter, S.R. (1990) Large-scale perturbations: opportunities for innovation. *Ecology* **71**, 2038–2043.

Chambers, J.M., Cleveland, W.S., Kleiner, B. & Tukey, P.A. (1983) *Graphical Methods for Data Analysis*. Wadsworth International Group, Belmont, CA.

Chapal, S.E. & Edwards, D. (1994) Automated smoothing techniques for visualization and quality control of long-term environmental data. In: *Environmental Information Management and Analysis: Ecosystem to Global Scales*. (eds W.K. Michener, J.W. Brunt & S.G. Stafford), pp. 141–158. Taylor & Francis Ltd., London.

Chatfield, C. (1984) *The Analysis of Time Series*. Chapman & Hall, London.

Cleveland, W.S. (1985) *The Elements of Graphing Data*. Wadsworth Advanced Books and Software, Monterey, CA.

Cleveland, W.S. & Devlin, S.J. (1988) Locally-weighted regression: an approach to regression analysis by local fitting. *Journal of the American Statistical Association* **83**, 596–610.

Cloern, J.E. & Jassby, A.D. (1995) Year-to-year fluctuation of the spring phytoplankton bloom in south San Francisco Bay: an example of ecological variability at the land-sea interface. In: *Ecological Time Series*. (eds T.M. Powell & J.H. Steele), pp. 139–149. Chapman & Hall, New York.

Conover, W.J. (1998) *Practical Nonparametric Statistics*. 3rd edn. John Wiley & Sons, Inc. New York.

Cressie, N. (1991) *Statistics for Spatial Data*. John Wiley & Sons, Inc. New York.

Debevoise, T. & Debevoise, N. (1999) *The Data Warehouse Method: Integrated Data Warehouse Support Environments*. Prentice Hall, Upper Saddle River, NJ.

Digby, P.G.N. & Kempton, R.A. (1987) *Multivariate Analysis of Ecological Communities.* Chapman & Hall, London.
Dobson, A.J. (1990) *An Introduction to Generalized Linear Models.* Chapman & Hall, London.
Draper, N. & Smith, H. (1981) *Applied Regression Analysis.* 2nd edn. John Wiley, New York.
Ellison, A.M. (1993) Exploratory data analysis and graphic display. In: *Design and Analysis of Ecological Experiments.* (eds S.M. Scheiner & J. Gurevitch), pp. 14–45. Chapman & Hall, New York.
Fayyad, U., Haussler, D. & Stolorz, P. (1996a) Mining scientific data. *Communications of the ACM* **39**, 51–57.
Fayyad, U.M., Piatetsky-Shapiro, G., Smyth, P. & Uthurusamy, R. (eds) (1996b) *Advances in Knowledge Discovery and Data Mining.* AAAI/MIT Press, Cambridge, MA.
Flohr, U. (1997) OLAP by Web. *Byte* **September**, 81–84.
Fortin, M. & Gurevitch, J. (1993) Mantel tests: spatial structure in field experiments. In: *Design and Analysis of Ecological Experiments.* (eds S.M. Scheiner & J. Gurevitch), pp. 342–359. Chapman & Hall, New York.
Frawley, W., Piatetsky-Shapiro, G. & Matheus, C. (1992) Knowledge discovery in databases: an overview. *AI Magazine* **Fall**, 213–228.
Gallagher, R.S. (1995) *Computer Visualization: Graphics Techniques for Scientific and Engineering Analysis.* CRC Press, Boca Raton, FL.
Gauch, H.G., Jr. (1982) *Multivariate Analysis in Community Ecology.* Cambridge University Press, London.
Goldberg, D.E. & Scheiner, S.M. (1993) ANOVA and ANCOVA: field competition experiments. In: *Design and Analysis of Ecological Experiments.* (eds S.M. Scheiner & J. Gurevitch), pp. 69–93. Chapman & Hall, New York.
Green, R.H. (1979) *Sampling Design and Statistical Methods for Environmental Biologists.* John Wiley & Sons, Inc., New York.
Gurevitch, J. & Hedges, L.V. (1993) Meta-analysis: combining the results of independent experiments. In: *Design and Analysis of Ecological Experiments* (eds S.M. Scheiner & J. Gurevitch), pp. 378–398. Chapman & Hall, New York.
Gurevitch, J., Morrow, L.L., Wallace, A. & Walsh, J.S. (1992) A meta-analysis of field experiments on competition. *American Naturalist* **140**, 539–572.
Haas, T. (1995) Local prediction of a spatio-temporal process with an application to wet sulfate deposition. *Journal of the American Statistical Association* **90**, 1189–1199.
Helly, J.J. (1998) Visualization of ecological and environmental data. In: *Data and Information Management in the Ecological Sciences: A Resource Guide* (eds W.K. Michener, J.H. Porter & S.G. Stafford), pp. 89–94, 134–137 (colour plates). LTER Network Office, University of New Mexico, Albuquerque, NM.
Hill, M.O. (1979a) *DECORANA—A FORTRAN Program for Detrended Correspondence Analysis and Reciprocal Averaging.* Cornell University, Ithaca, NY.
Hill, M.O. (1979b) *TWINSPAN—A FORTRAN Program for Arranging Multivariate Data in an Ordered Two-way Table by Classification of Individuals and Attributes.* Cornell University, Ithaca, NY.
Hill, M.O. & Gauch, H.G. (1980) Detrended correspondence analysis, an improved ordination technique. *Vegetatio* **42**, 47–58.
Hoaglin, D.C., Mosteller, F. & Tukey, J.W. (1983) *Understanding Robust and Exploratory Data Analysis.* John Wiley & Sons, Inc., New York.
Hollander, M. & Wolfe, D.A. (1998) *Nonparametric Statistical Methods.* 2nd edn. John Wiley & Sons, Inc., New York.
Hosmer, D.W. & Lemeshow, S. (1989) *Applied Logistic Regression.* John Wiley & Sons, Inc., New York.
Høst, G., Omre, H. & Switzer, P. (1995) Spatial interpolation errors for monitoring data. *Journal of the American Statistical Association* **90**, 853–861.
Ingersoll, R.C., Seastedt, T.R. & Hartman, M. (1997) A model information management system for ecological research. *BioScience* **47**, 310–316.

Inmon, W.H. (1992) *Building the Data Warehouse*. John Wiley & Sons, Inc., New York.

Jassby, A.D. & Powell, T.M. (1990) Detecting changes in ecological time series. *Ecology* **71**, 2044–2052.

Jensen, J.R. (1996) *Introductory Digital Image Processing: A Remote Sensing Perspective*, 2nd edn. Prentice Hall, Upper Saddle River, NJ.

Johnston, C.A. (1998) *Geographic Information Systems in Ecology*. Blackwell Science, Oxford, UK.

Jongman, R.H.G., ter Braak, C.J.F. & van Tongeren, O.F.R. (1987) *Data Analysis in Community and Landscape Ecology*. Center for Agricultural Publishing and Documentation, Wageningen, The Netherlands.

Legendre, P. (1993) Spatial autocorrelation: trouble or new paradigm? *Ecology* **74**, 1659–1673.

Ludwig, J.A. & Reynolds, J.F. (1988) *Statistical Ecology*. John Wiley & Sons, Inc., New York.

Manly, B.F.J. (1994) *Multivariate Statistical Methods: A Primer*. 2nd edn. Chapman & Hall, London.

McCullagh, P. & Nelder, J.A. (1989) *Generalized Linear Models*. 2nd edn. Chapman & Hall, London.

Michener, W.K. (1997) Quantitatively evaluating restoration experiments: research design, statistical analysis, and data management considerations. *Restoration Ecology* **5**, 324–337.

Mitchell, T.M. (1997) *Machine Learning*. McGraw-Hill, New York.

Myers, R.H. (1986) *Classical and Modern Regression with Applications*. Duxbury Press, Boston, MA.

Neter, J., Kutner, M.H., Nachtsheim, C.J. & Wasserman, W. (1996) *Applied Linear Statistical Models*. 4th edn. Irwin, Chicago, IL.

Nielson, G.M., Hagen, H. & Muller, H. (1997) *Scientific Visualization: Overviews, Methodologies, and Techniques*. IEEE Computer Society, Los Alamitos, CA.

Piegorsch, W.W., Smith, E.P., Edwards, D. & Smith, R.L. (1998) Statistical advances in environmental science. *Statistical Science* **13**, 186–208.

Potvin, C. (1993) ANOVA: experiments in controlled environments. In: *Design and Analysis of Ecological Experiments*. (eds S.M. Scheiner & J. Gurevitch), pp. 46–48. Chapman & Hall, New York.

Rasmussen, P.W., Heisey, D.M., Nordheim, E.V. & Frost, T.M. (1993) Time-series intervention analysis: unreplicated large-scale experiments. In: *Design and Analysis of Ecological Experiments*. (eds S.M. Scheiner & J. Gurevitch), pp. 138–158. Chapman & Hall, New York.

Scheiner, S.M. (1993) MANOVA: multiple response variables and multispecies interactions. In: *Design and Analysis of Ecological Experiments*. (eds S.M. Scheiner & J. Gurevitch), pp. 94–112. Chapman & Hall, New York.

Schneider, D.C. (1994) *Quantitative Ecology*. Academic Press, San Diego, CA.

Shaw, R.G. & Mitchell-Olds, T. (1993) ANOVA for unbalanced data: an overview. *Ecology* **74**, 1638–1645.

Sokal, R.R. & Rohlf, F.J. (1995) *Biometry*. W.H. Freeman and Company, New York.

Sprent, P. (1993) *Applied Nonparametric Statistical Methods*. Chapman & Hall, London.

Stewart-Oaten, A., Murdoch, W.M. & Parker, K.R. (1986) Environmental impact assessment: 'pseudoreplication' in time? *Ecology* **67**, 929–940.

Stewart-Oaten, A., Bence, J.R. & Osenberg, C.W. (1992) Assessing effects of unreplicated perturbations: no simple solutions. *Ecology* **73**, 1396–1404.

Stockwell, J.D., Burkey, T.V., Cazelles, B., Menard, F., Ascioti, F.A., Himschoot, P. & Wu, J. (1995) Detecting periodicity in quantitative versus semi-quantitative time series. In: *Ecological Time Series*. (eds T.M. Powell & J.H. Steele), pp. 99–115. Chapman & Hall, New York.

Strebel, D.E., Meeson, B.W. & Nelson, A.K. (1994) Scientific information systems: a conceptual framework. In: *Environmental Information Management and Analysis: Ecosystem to Global Scales*. (eds W.K. Michener, J.W. Brunt & S.G. Stafford), pp. 59–85. Taylor and Francis, Ltd., London.

ter Braak, C.J.F. (1988) *CANOCO—a FORTRAN Program for Canonical Community Ordination by (Partial) (Detrended) (Canonical) Correspondence Analysis, Principal Components Analysis and Redundancy Analysis, Ver. 2.1.* Rep. LWA-88-02, Agricultural Mathematics Group, Wageningen, The Netherlands.
Thierauf, R.J. (1997) *On-Line Analytical Processing Systems for Business.* Quorum Books, Westport, CT.
Thomsen, E. (1997) *OLAP Solutions: Building Multidimensional Information Systems.* John Wiley & Sons, Inc., New York.
Trexler, J.C. & Travis, J. (1993) Nontraditional regression analyses. *Ecology* **74**, 1629–1637.
Tufte, E.R. (1983) *The Visual Display of Quantitative Information.* Graphics Press, Cheshire, UK.
Tufte, E.R. (1990) *Envisioning Information.* Graphics Press, Cheshire, UK.
Tufte, E.R. & Krasny, D. (1997) *Visual Explanations: Images and Quantities, Evidence and Narrative.* Graphics Press, Cheshire, UK.
Tukey, J.W. (1977) *Exploratory Data Analysis.* Addison-Wesley, Reading, MA.
Underwood, A.J. (1991) Beyond BACI: experimental designs for detecting human environmental impacts on temporal variations in natural populations. *Australian Journal of Marine and Freshwater Research* **42**, 569–587.
Underwood, A.J. (1992) Beyond BACI: the detection of environmental impacts on populations in the real, but variable, world. *Journal of Experimental Marine Biology and Ecology* **161**, 145–178.
Underwood, A.J. (1994) On beyond BACI: sampling designs that might reliably detect environmental disturbances. *Ecological Applications* **4**, 3–15.
Underwood, A.J. (1998) Design, implementation, and analysis of ecological and environmental experiments. In: *Experimental Ecology: Issues and Perspectives.* (eds W.J. Resetarits, Jr. & J. Bernardo), pp. 325–349. Oxford University Press, New York.
Ver Hoef, J.M. & Cressie, N. (1993) Spatial statistics: analysis of field experiments. In: *Design and Analysis of Ecological Experiments* (eds S.M. Scheiner & J. Gurevitch), pp. 319–341. Chapman & Hall, New York.
von Ende, C.N. (1993) Repeated-measures analysis: growth and other time-dependent measures. In: *Design and Analysis of Ecological Experiments.* (eds S.M. Scheiner & J. Gurevitch), pp. 113–137. Chapman & Hall, New York.
Wahba, G. (1990) Spline functions for observational data. *CBMS-NSF Regional Conference Series, SIAM*, Philadelphia, PA.
Wiens, J.A. & Parker, K.R. (1995) Analyzing the effects of accidental environmental impacts: approaches and assumptions. *Ecological Applications* **5**, 1069–1083.
Winer, B.J., Brown, D.R. & Michels, K.M. (1991) *Statistical Principles in Experimental Design.* 3rd edn. McGraw-Hill, New York.
Zar, J.H. (1996) *Biostatistical Analysis.* 3rd edn. Prentice-Hall, Upper Saddle River, NJ.

CHAPTER 8

Ecological Knowledge and Future Data Challenges

WILLIAM K. MICHENER

8.1 Introduction

From a scientific standpoint, the intrinsic value of data is directly related to our ability to gain a higher level of understanding from those data; that is, the incipient information content of the data. Our knowledge base develops and evolves from the search for and identification of general patterns in ecology, patterns that often only become apparent after the results of numerous studies are examined. Such knowledge discovery can be technologically challenging for several reasons. First, significant attention must often be paid to managing and processing extremely large amounts of ecological data. Our ability to collect and store data can far exceed our ability to understand massive volumes of data. Consequently, ecologists must often work with some abstraction of the raw data, such as higher level information that emerges after substantial data processing (see Chapters 2, 3 & 7). Second, many diverse and sophisticated analytical tools are used to translate ecological data into the information that can facilitate interpretation and guide decision-making. These tools, many of which were identified in Chapter 7, require considerable time and effort to master, employ and interpret the results from their application. Finally, ecologists must frequently consolidate disparate data sets, re-examine previously collected data in light of newly emerging results and synthesize information and results from many previous studies. Such high level synthesis rarely occurs due to time and funding constraints, as well as the inherent difficulties in accessing the requisite data and information. This paucity of high level syntheses was one of the prime motivations for the establishment of the US National Center for Ecological Analysis and Synthesis (http://www.nceas.ucsb.edu).

The mere existence of information and the identification of general patterns in ecology do not necessarily mean that new information and knowledge will be put to use outside their role in scientific inquiry. For instance, special effort must often be put forth to insure that science overlaps with policy and resource management arenas. Many complex and important research questions, such as those related to sustainable production, protection of plant and animal diversity, and global climate and land use change, will guide ecology in the future. The importance of these issues to society, and the immediacy of the results of scientific studies, will require new and different approaches to science and technology.

Scientific and technological challenges and opportunities related to ecological knowledge discovery are the focus in this final chapter. First, ecological knowledge and several different approaches to science that may facilitate the acquisition and application of such knowledge to resource management, policy development, and decision-making are discussed. Second, technological challenges associated with developing the future ecological knowledge base are presented. Finally, several actions that can be taken now to facilitate management and translation of data into information and knowledge are recommended.

8.2 Ecological knowledge

Data consist entirely of characters and numbers that have little or no intrinsic meaning. Information, on the other hand, is a higher level representation of data where the data have been given form or character and confer meaning. Knowledge, then, is the understanding gained through discovery, perception and erudition of information.

Results from a single ecological study can seldom be equated with knowledge. Rather, knowledge consists of the general patterns that are perceived through the close study of information from many studies, studies that ideally represent similar tests repeated in different places and times. Underwood (1998, p. 343) emphasizes that 'a criterion of repeatability is a widespread safeguard against incautious belief in erroneous experimental results in many sciences. . . . When very different conditions reveal the same processes operating at similar magnitudes, there is more substance to our claim to understand what is happening in the world at large.' He further suggests that generality relies more on repeating small experiments in many different places and at many different times than in conducting a single statistically-powerful (low probability of Type I error) experiment. Attention to good data management practices (e.g. as discussed in Chapters 2–6) can greatly facilitate our efforts to repeat experiments over time, as well as synthesize data from many different studies for comparative analyses. The general ecological patterns that emerge from such efforts form the foundation for our knowledge. This continually evolving and expanding ecological knowledge base serves as a springboard for new ideas that lead to further scientific endeavours.

From a scientific standpoint, new pieces of knowledge gain value as they incrementally lead to additional searches for general patterns and subsequent scientific and technological advancements. Such new knowledge is typically presented at meetings and in traditional publication outlets that are familiar to all ecologists. Another, perhaps more pragmatic, 'litmus test' for ecological knowledge is the extent to which the understanding gained from scientific discovery is utilized for decision-making, resource management and policy development. New approaches for discovery and application of ecological knowledge will be explored in more detail in the following discussion.

The existence of new ecological knowledge does not necessarily translate into better decision-making, resource management and policy development. Clearly, educators, scientific societies and information-disseminating groups all play an important role in transporting ecological knowledge to the public. Baron and Galvin (1990) suggest that scientists also need to participate more directly in the active transport of ecological principles to groups such as policy makers by reporting research findings in policy journals using non-technical language. They also recognize, however, that these communication activities are dependent upon support and recognition of their value by peers and employing institutions.

In addition to after-the-fact communication of scientific findings, there are several other ways that ecological knowledge can be drawn more directly into the resource management and decision-making arenas. Three examples include developing science–management partnerships, participating in adaptive management and melding new ecological knowledge with the fortuitous opening of policy windows. These examples emphasize the importance of timely and effective communication and sharing of data and information among scientists, resource managers and policymakers.

8.2.1 Science–management partnering

As pointed out in Chapter 1, ecology is a relatively young discipline and many, if not most, of the mechanisms that underlie the structure and function of ecological systems remain to be unravelled. Consequently, the information and knowledge base available to guide resource management is often incomplete or inadequate. This situation may be further exacerbated if there is a disconnection (i.e. lack of timely communication) between science and resource management. One potential solution lies in the development of science–management partnerships.

Science–management partnering (Fig. 8.1) entails direct collaboration between ecologists and resource managers. 'Resource managers' is a broad category and includes individuals associated with industry, government and non-governmental organizations. The science–management partnering process consists of several steps. First, one or more questions emerge that are directly relevant to resource management. Scientists, often in conjunction with their management counterparts, rephrase the questions as one or multiple, unique and testable hypotheses. Examination and explicit statement of presuppositions, models, multiple working hypotheses and other research

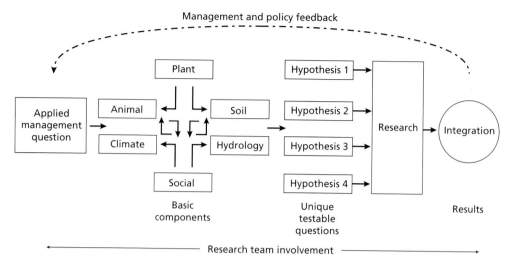

Fig. 8.1 Illustration of science–management partnering, a process whereby an interdisciplinary team of scientists collaborates with resource managers to address applied and basic questions. The transformation of data into information and knowledge is often facilitated in such endeavours. (Adapted from a slide made available by courtesy of Stephen Demarais, Department of Wildlife and Fisheries, Mississippi State University, Mississippi.)

activities listed in Chapter 1 should be integral to the process. It is especially important that the depth and breadth of the management 'problem' be proportional to the interdisciplinary nature of the research team (e.g. Caldwell 1996; Canter 1996). Studies are then designed to address the individual hypotheses. Results of the studies are then examined, integrated and synthesized by the science–management team into knowledge that then serves as the foundation for management and policy input. Finally, the management and policy input frequently leads to other questions that require additional research.

Science–management partnering is, of course, not a panacea. Such partnerships only work where there is good communication, when both sides contribute to the process, and when both sides gain from the collaboration. An example of an effective partnership might be the case where a research team receives funding from the resource management organization to address a series of questions that have both practical (applied) and basic importance. The research team may benefit enormously in situations where the resource managers are able to apply and replicate broad-scale treatments that are outside normal funding channels. The resource management organization may likewise benefit from the new insights gained from the research endeavour. Both sides may also benefit simply from the exposure to different mindsets, approaches and tool skills.

8.2.2 Adaptive management

Adaptive management is a systematic approach to improving management by learning from the results of management interventions. When active adaptive management is practiced, these management interventions are designed as experiments to test alternative ecological hypotheses, compare results to projected outcomes and apply new information and knowledge to developing more effective management strategies. Active adaptive management is especially useful when there is considerable uncertainty about the potential consequences of management practices. Passive approaches to adaptive management are based on the assumption that one particular management intervention is probably best, and involve implementation followed by monitoring and evaluation.

Adaptive management differs from conventional resource management and science–management partnering (Section 8.2.1) in several ways. First, conventional resource management ranges from cookbook to trial-and-error approaches. The former approach assumes a high degree of certainty in the outcome of management interventions, whereas seemingly overwhelming levels of uncertainty may lead to the latter approach. Trial-and-error may, in fact, lead to the desired outcome, but the lack of systematic monitoring and evaluation reduces the likelihood that the best management intervention can be identified. Second, science–management partnering is focused on employing the scientific method in the context of a single experiment to arrive at knowledge that may have both scientific and management implications. Adaptive management, on the other hand, has management as its central focus. Consequently, managers are integral to adaptive management and often assume leadership responsibilities. Furthermore, the focus in adaptive management is on monitoring and evaluating the success of the intervention(s) (i.e. the management experiment) on the system as a whole, and may involve monitoring only a few key parameters. Although science–management partnering is similar in many respects, the focus is often on developing a more mechanistic understanding of why something worked or didn't work, requiring comprehensive monitoring of many parameters.

The steps involved in adaptive management, as actively practiced, are shown in Fig. 8.2. First, the management problem must be clearly articulated. Second, the problem is then decomposed into a set of alternative hypotheses about how the ecosystem will respond to different interventions. Third, the interventions are carried out, ideally with suitable replication. Fourth, a limited number of key response variables are monitored over appropriate temporal and spatial scales. Finally, results of analyses provide feedback to improve the management knowledge base and define further adaptive management experiments. The entire process may be viewed as an upward spiral

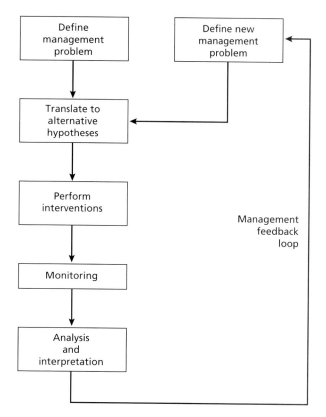

Fig. 8.2 Steps involved in adaptive management.

where feedback continues to improve management, leading to more refined hypothesis testing, *ad infinitum*. Excellent references for adaptive management include: Holling (1978); Walters (1986); and Taylor *et al.* (1997). In addition, Lee (1993) provides a very readable first-hand account of adaptive management in the Pacific Northwest of the United States.

8.2.3 Policy windows and scientific opportunities

Kingdon (1995) provides a model for how issues become subject to policy action. Essentially, the likelihood for action on a particular policy issue is a function of three interrelated events or processes (Fig. 8.3). First, there must be a problem or issue that occupies the attention of individuals associated with government at a particular time. Second, one or more policy alternatives must be articulated. Policy articulation relies upon the accumulated knowledge base that is brought to bear on an issue. Often, specialists with different perspectives will articulate very different policy options. Third, a political climate

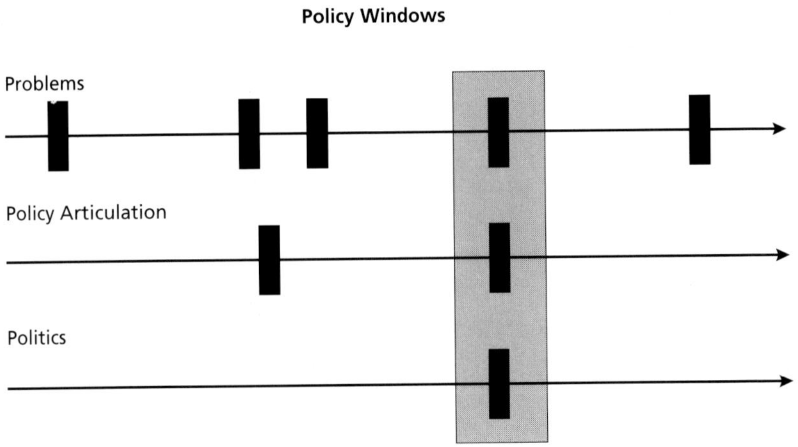

Policy Windows

Fig. 8.3 Opening of a policy window, a period when there is increased likelihood of authoritative action on an environmental issue that only arises when problems, policy alternatives and a favourable political climate converge. (From *BioScience* 48 (September 1998), page 767 (© 1998 American Institute of Biological Sciences) by permission of Richard A. Haeuber & William K. Michener.)

(e.g. public opinion, changes in administration, special interest group campaigns) that is favourable to change must exist.

The opening of a policy window occurs when there is a convergence of problems, policy alternatives and political climate (Fig. 8.3). Policy windows are critical periods because they are short-lived, occur infrequently, are surprising and often occur fortuitously. During the opening of a policy window, policy options are more likely to receive authoritative action. Haeuber and Michener (1998), for example, discuss the history of water management policy in the United States in relation to policy windows. Birkland (1997), similarly, relates natural disasters to policy windows.

In addition to problems, policy alternatives and political climate, there are at least two other 'Ps' that could be included as prerequisites for the opening of a policy window: procedure and persistence (Betsy Cody, personal communication). For instance, if a legislative window opens, an actionable policy may languish if correct procedures (e.g. committee assignment, scheduling and actions) are not followed. Second, persistence on the part of one or more 'champions' may be necessary to shepherd a policy option through the sometimes convoluted path to passage. Persistence is especially important for passage of solutions to environmental problems that must compete with financial, economic, military, bureaucratic, ideological and other considerations (Lemons 1996).

In the context of ecological knowledge, there are at least two take-home messages. First, the mere existence of ecological knowledge does not guarantee that it will be utilized for wise decision-making and policy articulation.

Ecologists can play an integral role in the policymaking process through timely communication of new discoveries and knowledge to the public and those involved in articulating policy alternatives. Second, because of the short-lived nature of policy windows, massive amounts of data may be of little use in articulating policy alternatives unless the data are understandable (i.e. associated with metadata) and readily accessible in a form that can be synthesized into higher level information. The various case studies included in *Bioregional Assessments* provide an in-depth view of how science is incorporated into policy and management (Johnson *et al.* 1999).

8.3 A Federated Information Infrastructure

Ecologists and resource managers are increasingly working at broader spatial and longer temporal scales. Attempts to scale research to the region, continent, and globe require unprecedented collaboration among scientists, data sharing across borders and ready access to high quality, well-documented data that have been preserved in data archives (i.e. the issues discussed in Chapters 1–6).

Following good or best data and information management practices facilitates research by individuals and institutions. But, good data management at the local level does not guarantee research success, particularly when the scope of inquiry is expanded to regional and global scales and data from many institutions are required to address specific questions. Attempts to broaden research efforts may fail because the data pertaining to environmental resources are incompatible, uncoordinated, too expensive, stored in isolated locations, inadequately protected or poorly documented. In essence, the requisite data are inaccessible or inadequate. In recognition of this problem in the United States, a President's Committee of Advisors on Science and Technology (PCAST) proposed development of a federated information infrastructure (FII) whereby terabytes of data from many different sources (satellite, field and laboratory) can be efficiently searched, data can be readily compiled in new ways for analysis and synthesis and the resulting information can be presented in understandable and useful formats (PCAST 1998; also see Robbins 1996).

FII is envisioned as a mechanism for providing seamless access to shared data and information resources. For instance, an ecologist interested in the effects of changing land use on primary productivity in a given area could easily and transparently access the requisite data layers (e.g. soil, land use and land cover, and up-to-date satellite imagery) from the appropriate governmental databases, regardless of where they physically reside. Database management operations (e.g. merging, subsetting) and sophisticated analyses could, for the most part, be performed at the click of a button. Although this example seems utopian at present, it is clear that significant improvements in the computational and communication infrastructure will be required to address many of these important environmental issues in a more comprehensive and timely fashion.

There are several technical (e.g. communication infrastructure, database interoperability, data archives), semantic (e.g. data compatibility, metadata, methods standardization) and social (e.g. scientific reward structure, cross-disciplinary communication) impediments to developing the federated information infrastructure that is envisioned (Stafford *et al.* 1994; Robbins 1996; PCAST 1998). Many of the challenges will require long-term funding, research and significant infrastructure improvements. Other challenges have reasonably straightforward solutions that can be implemented by simply following best data management practices.

8.3.1 FII technical challenges

Technical challenges include the need for an improved communication infrastructure that can handle massive bandwidth requirements and advanced network architectures. Database interoperability must be significantly improved. New and expanded data archives will be necessary to provide both secure storage and ready access to environmental data.

Many of the most difficult challenges that lie ahead relate to transforming data into information and, ultimately, knowledge. Knowledge about our environment entails the synthesis of data from many sources and typically requires that a scientist acquire, manage, manipulate, correlate, analyse and synthesize data from individual data sets, one at a time. The continuance of long-term monitoring programmes, coupled with significant improvements in sensors and data acquisition, have led to the current situation where many organizations now have an excess of data. Exponential increases in the size of data holdings coupled with the recent development of new environmental remote sensing technology (e.g. multispectral data at 1–3 m^2 spatial resolution) requires the development of new approaches to exploit these massive data sets. Specifically, we need tools to analyse and synthesize data quickly and to translate those data into useful information that can guide decision-making, policy formulation and future research.

Critical technological challenges include the need to develop:

1 extraction and analytical tools for correlating, manipulating, analysing and presenting distributed information (i.e. new analytical (statistical and modelling) techniques that work with multidimensional, large-volume data);

2 new quality assurance methods that 'correct' data errors with minimal human intervention;

3 metadata encoding routines to facilitate data mining of these massive data sets;

4 algorithms for analysis, change detection and visualization that scale to large, multi-temporal and multi-thematic databases.

8.3.2 FII semantic challenges

Semantic challenges encompass those factors that lead to difficulties in understanding and interpreting data. Ecological data are particularly complex, encompassing millions of different organisms and hundreds to thousands of different communities and ecosystems. Such data are collected by different countries, agencies, industries, academic institutions and individuals, all of which have different needs, views, requirements and skills. Furthermore, the data vary substantially in scale, precision, accuracy, type (text, measurements, images, sound, video) and volume (kilobytes to terabytes). Consequently, even within a single institution, data pertaining to one environmental parameter often cannot be compared with data on other parameters because of different data structures, scales of measurement, region of coverage, times data were collected and so on. Problems associated with different data collection protocols, data storage mechanisms and comprehensiveness of data documentation are compounded when more than one institution is involved.

Two specific actions that can minimize semantic conflicts include the development and adoption of metadata standards, as well as the standardization of data collection protocols where possible (see Chapters 1 & 5). Metadata provide critical information for expanding the scales at which ecologists work. For example, field validation data from multiple sites are frequently used to calibrate (or, in some cases, are merged with) remotely sensed data, thereby expanding the spatial domain from the site to broader scales. Cross-site comparative studies depend heavily upon the availability of sufficient metadata. For cross-site comparisons, it is especially important that both methods and instrumentation calibration and inter-calibration (measurements of similar parameters by different methods or instruments) be well documented to confirm data integrity, proper use of experimental methods and data acquisition.

Much of the *post hoc* effort that is devoted to managing (manipulating), merging and analysing data for parameters that were collected under different protocols can be reduced or eliminated when standard methods are employed a priori (see Chapter 1). Furthermore, development and adoption of standard data collection and management protocols reduces the amount of time and effort expended in developing metadata.

8.3.3 FII social challenges

Future research and resource management will require unprecedented collaboration among scientists from many disciplines, as well as data sharing across departmental, agency, academic and national borders. Success will, to a large degree, depend on the extent to which we alter existing scientific reward structures. Data sharing and collaboration are facilitated when the stakeholders perceive that there are real benefits in doing so (Porter & Callahan 1994). Thus, if the delivery of useful data products is part of an organization's objective, then those contributing to the development of the product (i.e. database) deserve credit for doing so. In essence, databases should be viewed as being synonymous to a publication and should be considered in personnel review and promotion procedures (also see Chapter 6).

The lack of effective cross-disciplinary communication, data sharing and collaboration often impedes attempts to broaden the scope of our scientific efforts. Improved communication is key to resolving conflicts among participants with different training, vocabularies and worldviews. Where possible, it is beneficial to build upon past successes and plan early for collaborative efforts.

8.4 Where do we go from here?

A primary challenge in designing a research project lies in striking a rational balance between economic feasibility and the data scale(s) and volume that are optimally required to meet scientific objectives. High quality, well documented, securely preserved and accessible data are essential for addressing most long-term and broad-scale environmental problems. Access to high quality data requires a strong commitment to implementation of effective information management procedures. The absence of such procedures impairs our ability to use data over long periods. Adherence to recommended data management practices, especially the development of comprehensive metadata and the submission of both data and metadata to data archives, greatly slows the progression of data entropy (see Chapters 2, 5 & 6).

Solutions to many of the problems inherent in establishing the FII will require a substantial infusion of money, personnel, creative thought and technology by businesses, research and resource management organizations and nations. However, there are at least seven steps related to information management that can be taken to facilitate the more responsive science and resource management needed now and for the future.

1   Allocate a reasonable percentage of research funding for management of data and information generated by the research. In most organizations, data management is seriously under-funded, resulting in data losses and delays in translating data to information.
2   Develop and adhere to data and metadata standards and best use protocols (e.g. Chapters 2–6; Michener *et al*. 1994, 1997, 1998).
3   Provide funding for data rejuvenation (e.g. adding Global Positioning

System fixes, i.e. latitude and longitude, to field sites) and rescue (e.g. convert paper records to digital format) to halt further data entropy.

4   Routinely evaluate data utility, research objectives and management needs, and re-establish priorities. Use this information to revise long-term sampling programmes (e.g. reduce effort in certain areas, add new parameters) and to streamline data capture.

5   Coordinate software and systems development and purchases with other agencies or departments to eliminate duplication of effort and reduce expenditures (i.e. take advantage of economies of scale).

6   Cooperate with other agencies, scientists and the private sector to establish and adopt data and metadata standards, authority files and thesauri for data, as well as support data archives.

7   Establish formalized education outreach and training in ecological informatics (i.e. the design, management and processing of ecological data) to develop the scientific resource base for the future.

8.5 References

Baron, J. & Galvin, K.A. (1990) Future directions of ecosystem science. *BioScience* **40**, 640–642.

Birkland, T.A. (1997) *After Disaster: Agenda Setting, Public Policy, and Focusing Events*. Georgetown University Press, Washington, DC.

Caldwell, L.K. (1996) Science assumptions and misplaced certainty in natural resources and environmental problem solving. In: *Scientific Uncertainty and Environmental Problem Solving*. (ed. J. Lemons), pp. 394–421. Blackwell Science, Cambridge, MA.

Canter, L.W. (1996) Scientific uncertainty and water resources management. In: *Scientific Uncertainty and Environmental Problem Solving*. (ed. J. Lemons), pp. 264–297. Blackwell Science, Cambridge, MA.

Haeuber, R.A. & Michener, W.K. (1998) Policy implications of recent natural and managed floods. *BioScience* **48**, 765–772.

Holling, C.S. (1978) *Adaptive Environmental Assessment and Management*. John Wiley & Sons, Inc., London.

Johnson, K.N., Swanson, F., Herring, M. & Greene, S. (1999) *Bioregional Assessments*. Island Press, Washington, DC.

Kingdon, J.W. (1995) *Agendas, Alternatives, and Public Policies*, 2nd edn. Harper Collins, New York.

Lee, K.N. (1993) *Compass and Gyroscope*. Island Press, Washington, DC.

Lemons, J. (1996) Introduction. In: *Scientific Uncertainty and Environmental Problem Solving*. (ed. J. Lemons), pp. 1–11. Blackwell Science, Cambridge, MA.

Michener, W.K., Brunt, J.W. & Stafford, S.G. (eds) (1994) *Environmental Information Management and Analysis: Ecosystem to Global Scales*. Taylor and Francis, Ltd., London.

Michener, W.K., Brunt, J.W., Helly, J., Kirchner, T.B. & Stafford, S.G. (1997) Non-geospatial metadata for the ecological sciences. *Ecological Applications* **7**, 330–342.

Michener, W.K., Porter, J.H. & Stafford, S.G. (eds) (1998) *Data and Information Management in the Ecological Sciences: A Resource Guide*. University of New Mexico, Albuquerque, NM. (Available through http://www.lternet.edu/ecoinformatics/guide/frame.htm)

PCAST (President's Committee of Advisors on Science and Technology) (1998) *Teaming with Life: Investing in Science to Understand and Use America's Living Capital*. Office of Science and Technology Policy, Washington, DC (Available through http://www.whitehouse.gov/)

Porter, J.H. & Callahan, J.T. (1994) Circumventing a dilemma: historical approaches to data sharing in ecological research. In: *Environmental Information Management and Analysis: Ecosystem to Global Scales.* (eds W.K. Michener, J.W. Brunt, & S.G. Stafford), pp. 193–203. Taylor and Francis, Ltd., London.

Robbins, R.J. (1996) Bioinformatics: essential infrastructure for global biology. *Journal of Computational Biology* **3**, 465–478.

Stafford, S.G., Brunt, J.W. & Michener, W.K. (1994) Integration of scientific information management and environmental research. In: *Environmental Information Management and Analysis: Ecosystem to Global Scales.* (eds W.K. Michener, J.W. Brunt & S.G. Stafford), pp. 3–19. Taylor and Francis, Ltd., London.

Taylor, B., Kremsater, L. & Ellis, R. *Adaptive Management of Forests in British Columbia.* Ministry of Forests, Forest Practices Branch, Victoria, British Columbia.

Underwood, A.J. (1998) Design, implementation, and analysis of ecological and environmental experiments. In: *Experimental Ecology: Issues and Perspectives.* (eds W.J. Resetarits, Jr. & J. Bernardo), pp. 325–349. Oxford University Press, New York.

Walters, C.J. *Adaptive Management of Renewable Resources.* Macmillan, New York.

Index

Please note: page numbers in *italics* refer to figures; those in **bold** refer to tables.

access to data, 28
   access policy, 42
   data archives, 41, 125–6, 137–8
accessibility descriptors, 97, **99**
acquisition of data, 29, 34–6
adaptive management, 166–7, *167*
administration of data, 29, 41–6
   personnel, 44–6
   requests handling, 44
   research support functions, 42–4, **43**
analysis of variance (ANOVA), 149
archive operation, 131–5, **132**, *133*
   data addition, 131
   data ingestion, 132
   data preservation, 132–5
   data review, 131
archiving data, 29, *39*, 40–1, 117–40
   access, 41
      immediate, 125–6
   computer technology, 138, **139**
   contributors, 123
   data delivery, 137–8
   data preparation, 128–31, **129**
      guidelines, 128–9
      identifiers, 130
      location ID, 130
      storage, 130
      temporal data, 130
      transport, 130–1
      unique occurrences, 129–30
   data rescue, 127
   data subsets, 137
   data transformation processes, 145
   definitions, 117–18, **119**
   expansion of archives, 139–40
   facilitation, 140
   files, 40–1
   formal archival data storage, 119–23
   funding, 123
   incentives, 123–4
   index system, 136
   locating data, 136–8
   practical problems, 123
   security, 133–4
   time frame for availability, 125–6
   user interface, 136–7
   user support, 138–9
ASCII files, 135
automated data acquisition systems, 36
autoregressive (AR) models, 151
autoregressive integrated moving average (ARIMA) models, 151
autoregressive moving average (ARMA) models, 151

back-up, 36
   archive data, 133–4
   restoration, 134
balance of nature paradigm, 4
Before-After-Control-Impact (BACI) studies, 150, 153, *154*
biochemical dynamics data archives, **120**, 120, 121
biodiversity data management, 64
BIOTA, 64
box-and-whisker plots, 146, *147*
Breeding Bird Survey, 121, **122**

C++, 57
Caenorhabditis elegans, 65, 66
canonical correspondence analysis, 150
Carbon Dioxide Information Analysis Center (CDIAC), 108, 119, **120**
Catalogue of Data Sources (CDS), 121, **122**
Caterpillar Hostplants Database, 127
causation, 9
climate data archives, 119
climate monitoring, 101
cluster analysis, 149, 150
collaboration, 126, 171–2
commercial software, 57, 58
comparative studies, 11–12
complexity, 7
   of data, 49, *50*
computer technology, 29
   data archives, 138, **139**
   system selection, 59–60
conceptual models, 4
   hypothesis generation, 5, *6*
   multiple working hypotheses, 7
confounding factors, 8, 142

175

INDEX

contamination of data, 36
  Grubbs' test, 77–8, 88
  outliers, 73
  prevention, 70, 71–3
Content Standards for Digital Geospatial Metadata, 96
CORBA, 60
correlation, 9
correspondence analysis, 149, 150
cross-site comparisons, 171

data analysis *see* data transformation
data design, 29, *30*, 30–4
  data set organization, 31–3
  data tables, *31*, 31, *32*
data documentation, *see* metadata
data entry errors, 36, 71
Data File Management System (DFMS), 32–3, *33*
data (information) manager, 45
data licenses, 44
data loggers, 36, 144
data loss, 133
data management, 3–4, 25–46, 143, 169, 172–3
  administrative procedures, 29, 41–6
  archival storage *see* archiving data
  basic principles, 25–6
  data acquisition, 29, 34–6
  data design, 29, 30–4
  documentation protocols, 29, 37–40
  flexibility, 25, 26
  funding, 172
  implementation, 28–9
    rationale, **28**
  inventory of existing data/resources, 29, 30
  metadata, 29, 37–40, *40*
  quality assurance, 29, 36–7, **38**
  role, 26–7, *27*
  scientist participants, 25, 26
  system components, 29
data manipulation, 33–4
data maturation, 37, 124
data mining, 156–7
data preservation, 132–5
  back-up, 133–4
  digital storage, 134–5, **135**
  electronic storage upgrades, 133, 134
  paper records, 133, 134
data processing, 143–5, *144*
  federated information infrastructure (FII), 170
  flow chart, *39*
data reduction, 143–4
data rescue, 127, 172–3
data reuse, 94, 95
  metadata, 111
    requirements, 107–8, 110
    structure, 105

data set
  descriptors, 97, **98**, **99**
  inventory, 29, 30
  organization, 31–3
data sharing, 94, 95, 118, **119**
  data archives, 117, 118
  facilitation, 140
  federated information infrastructure (FII), 169, 171–2
  incentives, 124
  investigator to investigator, 126
  open (secondary users), 126–7
  phases, 124–7, **125**
  time frame for immediate availability, 125–6
data storage formats, 104
data structural descriptors, 97, **99–100**
data tables, *31*, 31
  normalization, 31, 60
  relational attributes, 31, *32*
data transformation (analysis), 142–58
  data processing, 143–5, *144*
  graphical analysis, 145–9
  statistical analysis, 145, 149–56
  very large databases, 156–8
data volume, 49, *50*
data warehousing, 67, 110, 157–8
database management system (DBMS), 34, 53–5, **54**
  data archiving, 130
  data tables, 31
  data file management system, 33
  metadata, 105, 106
  software considerations, 57–8
  transferring field data, 37
  World Wide Web (WWW) server interface, 58–9
degradation of data, *93*, 93–4, 124–5
dendrochronological analyses, 11
descriptive statistics, 149–50
detrended correspondence analysis, 150
digital image processing, 145
digital storage, 134–5, **135**
Directory Interchange Format (DIF), 101, **102**, 106
diskettes, archive data delivery, 138
Distributed Active Archive Center for Biochemical Dynamics (ORNL DAAC), 119–20, **120**, 121, 127
Distributed Active Archive Centers, 64
distribution assumptions checking, 74
disturbance, 4
diversity indices, 149
double keying of data, 71
Dublin Core, 101, **102**

Earth Observing System Data and Information System (EOSDIS), 120
Earth Observing System (EOS), 120, 121

ecological knowledge, 163–4
    application, 164
    policymaking processes, 168–9
Ecological Monitoring and Assessment
    Network (EMAN) in Canada, **122**
ecological restoration, 13–14
electronic storage upgrades, 133, 134
EMBL, 65
entity-relationship (E-R) diagram, 61
Environmental Information Centre, **122**
environmental data archives, 121, **122**
    climate, 119
environmental sensors, 36
ER/Studio, 63
errors in data, 36, 71
ERWin, 63
European Environmental Agency Catalogue
    of Data Sources (CDS), 121, **122**
Excel, 34
experimental design, 7–9, *9*
    data tables, *31*, 31
exploratory data analysis, *146*, 146–7, *147, 148*
eXtensible Markup Language (XML), 106–7

federated information infrastructure (FII),
    169–72
    collaborative aspects, 171–2
    semantic conflicts, 171
    technical aspects, 170–1
file management system, 31, 33
file-system-based databases, **54**, 54–5, *55*
flux of nature paradigm, 4
Forest Inventory and Analysis Database,
    121, **122**
Foxpro, 33

GenBank database, 52, *53*, 53, 63–4, 65
generalized linear models (GLiMs), 156
geophysical data archives, 121
geophysical databases, 64
geostatistics, 155
GIS (geographic information systems), 34,
    49–50, 96, 104
Global Climate Observing System, 101, 103
Global Historical Climatology Network, 110
global terrestrial net primary productivity,
    127
GLS variogram method, 155
Government Information Locator Service
    (GILS), 101, **102**
graphical analyses, 145–9
Grubbs' test, 77–8, **78**, *79*, 88

hand-held computers, 35–6
herbarium specimens management, 64
Hierarchical Data Format (HDF), 135
hierarchical databases, **54**, *55*, 55
Human Genome Project, 52, *53*, 53, 65
human influences, 4

Hypertext Markup Language (HTML), 106
hypothesis generation, 5–6, *6*

illegal data filters, 71–3, **72**, *72*
impact level-by-time interaction analysis, 153
influential observations *see* leverage points
information entropy, *93*, 93–4, 124–5
instrument miscalibration, 89
integrated databases, 128
integrative research approaches, 60, 109–11
International Biological Program (IBP) data,
    133
International Soil Reference and
    Information Center (ISRIC), 121, **122**
intervention analysis, 151–2
Iterative Reweighted Least Squares (IRLS),
    89

JAVA, 57, 59

keywords, 132
knowledge discovery in databases (KDD),
    156–7
Kolmogorov-Smirnov test, 74
kriging, 155

large-scale studies, 8, 11–12
Latin square design, *9*
leverage points
    multiple linear regression, 84, 86, 88
    simple linear regression, 79–81, *80*
Linnean herbarium database, 121
literature review, 15
log-normally distributed data, 89
log-transformed data, 89
logistic regression, 155, 156
Long Term Ecological Research (LTER)
    network, 64, 121, **122**
long-term studies, 1–2, 8, *10*, 10–11, 14, 169
    archiving of data, 123
    data management, 26, 27, 143
    data rescue, 127
    data reuse, 95
    database data, 48, 50–1
    metadata, 94
Lotus 1-2-3, 34

MANOVA, 156
Mantel tests, 155
MEGRIN, 97
Mercury, 103
meta-analysis, 156
metadata, 29, 37–40, *40*, 49, 92–112
    archive data, 118
        review, 131
    benefits, 94–5
    content standards, 96–103, 171
        geospatial metadata, 96–7
        non-geospatial ecological metadata, 97
    practical limitations, 111–12

costs, 95, 107, 108, 112
database management systems, 58
degradation, *93*, 93–4
descriptors, 109
  generic formats for resources, 101, **102**, 103
  geospatial metadata, 96
  non-geospatial ecological metadata, 97, **98–100**, 101
  federated information infrastructure (FII), 170, 171
  implementation, 107–11
    needs assessment, 108–9
    pilot project, 109
    practical aspects, 108–9, **110**
  incentives, 125
  integrated storage, 40
  planning issues, 107–8
  software, 103–4
    development tools, 106–7
    specificity of requirements, 104
    Web sites, 103
  structure, *105*, 105–6
  unstructured, 104
  Web-based search and data retrieval systems, 103
Microsoft Windows NT, 60
modelling projects, 60–3, **61**, *61*, **62**, *62*, *63*, 109–11, 117
  availability of archived data, 127
moving average (MA) models, 150–1
MS SQL Server, 34
multi-disciplinary studies, *2*, 2–3, 48, 117
  wide databases, 64
multidimensional scaling, 150
multi-investigator projects, 126
multiple linear regression, 149
  outliers detection, 84–8
    diagnostic measures, 84
    example, *85*, 85–8, *86*, *88*
multiple logistic regression, 155
multiple working hypotheses, 7
multivariant analysis of variance (MANOVA), 149
MUSE, 64

NASA Earth Observing System (EOS), 120, 121
NASA Global Change Master Directory, 122
National Center for Ecological Analysis and Synthesis, 162
National Climatic Data Center (NCDC), 119, **120**
National Environmental Data Index (NEDI), 121–2, **122**
National Environmental Data Referral Service (NEDRES), **102**
National Environmental Satellite, Data and Information Service (NESDIS), 121, **122**

National Geophysical Data Center, 121, **122**
National Ocean Data Center, 121, **122**
National Oceanic and Atmospheric Administration (NOAA), 119, 121
National Soils Data Access Facility (NSDAF), 121, **122**
National Water Information System, 121, **122**
NBII MetaMaker, 103
NERC Data Centres Threatened and Endangered Species, **122**
Network Common Data Format (NetCDF), 106, 135
network data transfer, 138
network databases, **54**, 56
nonlinear regression, 156
nonparametric statistical approaches, 155
normal probability plots, 74–6, **75**, *75*, *76*
  non-normally distributed data, *76*, *77*
normalization, 61
Normalized Difference Vegetation Index, 145

Oak Ridge National Laboratory Distributed Active Archive Center (ORNL DAAC), 119–20, **120**
Object Data Base Connectivity (ODBC), 59
object-oriented databases, **54**, 57
  management systems (OODBMS), 40
online analytical processing (OLAP), 157–8
Oracle, 33
ordination techniques, 149, 150
orphaned data, 127
outliers detection, 73–6
  Grubbs' test, 77–8, **78**, *79*
  multiple linear regression, 84–8
  normal probability plots, 74–6, **75**, *75*, *76*, *77*
  robust regression models, 89, *90*
  simple linear regression, 80–1
ozone levels, 73

palaeoecological analyses, 11
paper records
  archive data, 133, 134
  data forms, 35
  unstructured metadata, 104
paradigms, 4
parameter numbers (thematic scale), *1*, 1, 15
PDB, 64
periodicity in time series, 151
pilot studies, 18
  metadata implementation, 109
point-of entry data quality checks, 35–6, 37
policy issues, 167–9
policy windows, *168*, 168
principle components analysis, 149, 150
principle coordinate analysis, 150

Printing using Error Recovery Method 1
  (PERM1), 134
probes, 36
protein sequence database, 64
protein structure database, 64
pseudoreplication, 8
public domain software, 57, 58

quadrat variance, 149
quality assurance (QA), 29, 35, 36–7, **38**,
  70–90, 108
  federated information infrastructure (FII),
    170
  graphical approaches, 147
  mechanisms, 37
  preventive approach, 70, 71–3
  user motivation, 70, 71
quality circles, 71
quality control (QC), 36–7, **38**, 108

radiometers, 36
randomized block design, 9
randomized experimental design, 9
raw data
  degradation, *93*, 93–4
  processing, 143, *144*
relational databases, **54**, *56*, 56, 58, 106
remote sensing, 96, 170, 171
repeated measures models, 156
replication, 8, 11, 14, 15, 142, 163
  standard methods, 16
research approaches, 7–14
research design, 1–18
  implementation issues, 14–18
research domain, 14–16
research origin descriptors, 97, **98–9**
retrieval fees, 44
retrospective studies, 11
Ribosomal Database Project, 65
robust regression techniques, 89, *90*, 151

S+ (Splus), 33, 89, *90*, 130
  normal probability plots construction, **75**,
    75, 75, *76*
sample size, 16
sampling design, 16
SAS, 33, 34, 72, 130
  normal probability plots construction, **75**,
    75, 75
SAS Insight, 147, *148*
satellite information data archives, 121
scalability, 53, 60
science-management partnership, 164–5, *165*
scientific databases, 48–67
  computer system selection, 59–60
  data diversity, 49–50
  data modelling, 60–3, **61**, *61*, **62**, *62*, *63*
  data structure, 52
  database systems, 53–5, **54**
  deep, **63**, 63

  development, 51–3
  funding, 66–7
  incentives for data contributors, 52, 65, 66
  interlinking resources, 60
  requirements, 49–51
  search capability, 52
  user community, 51, 64–5
  user diversity, 50, 51
  very large, 156–8
  wide, **63**, 63, 64
scientific visualization, 149
secondary data use, 117, *118*
  metadata requirement, 107–8, 110
  open data sharing, 126–7
  simulation models, 110–11
security needs of data archives, 134
short-term studies, 14
  integration into long-term studies, 94
simple linear regression, 149
  brain weight vs. body weight, 81–2, *82*, *83*
  leverage points, 79–80, *80*
  outliers, 79–83, *80*
    diagnostic measures, 80–1
    example, 81–3, *82*, *83*
simulation models, 12–13
  metadata, 110–11
software, 52, 57–8
  metadata, 103–4
  user base, 58
soil methods, 101
soils data archives, 121, **122**
space-for time substitution, 12
spatial analytical approaches, 154–5
spatial autocorrelation analysis, 155
spatial data metadata standards, 96–7
spatial scale, *1*, 1–2, 5, 14, 15, 169
spatial searching, 52, 53
split plot design, *9*
spreadsheet software, 34, 130
standard methods, 16–17, **17**, 50
statistical analysis, 145, 149–56
  conventional statistics, 149
  descriptive statistics, 149–50
  nonparametric approaches, 155
  spatial approaches, 154–5
  temporal approaches, 150–4
statistical power, 16
Structured Query Language (SQL), 57
succession, 4, 12
Swedish Museum of Natural History
  databases, 121, **122**
SWISS-PROT, 64
synthesis projects, 109–11, 117, 162
  availability of archived data, 127

tape media, 35
  archive data delivery, 138
tape recorders, 35
temporal data reduction, 143–4
temporal statistical analysis, 150–4

time scales, *1*, 1–2, 5, *10*, 14, 15
  simulation modelling, 13
time-series analysis, 150–3, *152*
twenty-year rule, 130–1

UNIX, 60, 66
unstructured metadata, 104

very large databases, 156–8

Warwick framework, 104
water resource data archives, 121

Web-based applications, 34, 58–9
  metadata, 103, 106
    implementation tools, 106–7
wetland mitigation projects, 14
within-class variance, 18
word processing formats, 130
Worm Community System, 65, 66
worm ontogeny databases, 65–6

X–Y scatter plots, *146*, 146

Z39.50, 60